FOR POLLUTION FIGHTERS ONLY

FOR POLLUTION FIGHTERS ONLY

Margaret O. Hyde

Illustrated by Don Lynch

McGRAW-HILL BOOK COMPANY

New York · St. Louis · San Francisco · Düsseldorf
Johannesburg · Kaula Lumpur · London · Mexico · Montreal
New Delhi · Panama · Rio de Janeiro · Singapore
Sydney · Toronto

OTHER BOOKS BY MARGARET O. HYDE

ANIMAL CLOCKS AND COMPASSES
ANIMALS IN SCIENCE
DRIVING TODAY AND TOMORROW, Rev. Ed.
EXPLORING EARTH AND SPACE, Fifth Ed.
FLIGHT TODAY AND TOMORROW, Third Ed.
FROM SUBMARINES TO SATELLITES
MEDICINE IN ACTION, Rev. Ed.
MIND DRUGS
MOLECULES TODAY AND TOMORROW
OFF INTO SPACE, Third Ed.
PLANTS TODAY AND TOMORROW
THIS CROWDED PLANET
YOUR BRAIN—MASTER COMPUTER
YOUR SKIN

With Bruce Hyde

ATOMS TODAY AND TOMORROW, Fourth Ed.

With Edwin Hyde

WHERE SPEED IS KING, Rev. Ed.

With Dr. Edward S. Marks

PSYCHOLOGY IN ACTION

Library of Congress Catalog Card Number: 77-151497
34567890 VBVB 79876543
07-031629-5
07-031628-7

Contents

The Author is grateful to the National Wildlife Federation for permission to reprint *How to Organize a Community Awareness Program*

To Seymour H. Cohen, M.D.

1 POLLUTION: CHALLENGE TO YOU

Someday, when you are returning from a trip in space to the beautiful, blue earth, a signal may flash to show that the air within your craft is becoming polluted. At once action will be taken to freshen the air.

Your spaceship is a cocoon in which all the things that support life and provide comfort are carefully watched and controlled. It is one kind of ecosystem or life-support system. The air that the passengers breathe is recycled from air that they breathed before. The water that they drink is recycled from water that they drank before and expelled as body waste. Materials are used again and again, for only a limited amount can be carried through space.

Back on Planet Earth people live in a sort of cocoon or spaceship that carries them around the sun. The whole earth is an ecosystem. The air that people

breathe is used again and again. The water that they drink was used by someone else before them. And the soil on which they live is limited in supply, just as are the water and the air.

Spaceships for travel away from the earth are small compared to the giant ecosystem which is the planet itself. On the earth there are ecosystems of various sizes. Some are fairly simple; others are very complicated. No matter what the size or kind or place of the ecosystem, it must provide certain things to support life. The quality of these things can make a great difference in the quality of the life which it supports.

Today many people are sounding the alarm button on Spaceship Earth. The air is becoming so polluted that it is causing serious diseases and numerous deaths among the passengers. Air pollution alone is one kind of pollution that snuffs out lives of the many suffering from respiratory ailments. People are calling attention also to the threats of hunger, of drowning in the world's wastes, of lack of privacy, of poisoned water, and of many other dangers.

Certainly man's recent voyage to the moon helped to make him realize that he traveled on a small spaceship that was dependent on a limited amount of air, water, and land. From the time of early man to the time of the astronauts, few men realized how closely all living things in the world are related to each other. Today many people, and especially young people, are looking at the whole world in an effort to learn about the life web that threads through the entire planet, a life web of different kinds of ecosystems such as forests, deserts, and Arctic tundra, the life web of each small community as a separate and as a connected part.

You are needed right now to help fight pollution. The world cannot wait for you to grow older. Millions of people are asking, "Can pollution be stopped? Do we have time to change the course of events so that man can survive on the earth?" Perhaps man can adapt to an environment of poorer quality. But must he? Part of your job is to create increased awareness, to get people to realize how serious the problem is. Another part is to help solve the problem. You can help to fight pollution in many ways. You can do it by looking, by learning, by experimenting with the relationships of living things in your own community. You can do it by helping to correct problems that cause pollution. You can help by your own knowledge and by spreading this knowledge to others so that each person can help in his own special way.

No one knows if the rivers will run clear again and the

sky will be free of smog. The important question is not the emotional one, "Can man survive?" Man must accept the challenge and act wisely. What can man do to keep his planet fit for living? How can he improve the quality of life? How can he reverse the trend to pollution?

Your job as a pollution fighter is an important one. This book will tell you how to help as an individual, or as part of a group. Just as the smog is caused by many individual particles, the work of individuals can and must create a tremendous change. Millions of young people have accepted the challenge of seeking solutions for the problems of the environment of Planet Earth. As an individual and as one of a group, you are needed. On Earth Day, April 22, 1970, many people began to consider their own individual roles in healing the sick environment in which they found themselves. Thousands of citizen action groups sprouted as a result of the efforts of young people. Students are writing new study plans and exploring some old ones, that were once considered dull, with the greatest enthusiasm. The young are becoming the teachers, for it is their world that is most endangered. Learning about environment and ecology and what can be done to decrease pollution is a complicated matter. But long after Earth Day, the learning and the action must go on.

Almost all pollution problems are connected to each other, and almost all pollution problems are connected to people. If you can help in one way, you are probably helping in many. Unfortunately, the tremendously complex connections in the problems of today's environment are not recognized by many who are looking for simple and instant solutions. The interdependence of man and his environment cuts across the entire fabric of life. Correcting the problems of environment demands a revolution in the way men approach problems and make decisions. Man cannot win the battle for a natural environment of quality by suddenly declaring that we should live in harmony with nature. Talking about the junk piles of automobiles, doing away with pesticides, talking about tanker oil spills and the invasion of the ocean by chemicals (including as much as 4,000 tons of mercury each year) will not bring a pollution-free tomorrow.

Even the pollution fighter contributes to pollution. There is no instant end once one recognizes the seriousness of the situation.

Suppose you decided to stop polluting the world. Turn off your oil burner; box your trash; don't ride in an airplane, an automobile, a bus, a train, a motorcycle, or anything that is powered by energy other than your own; refuse the delivery of anything that comes to your house on any of these vehicles; do not use anything that requires pollution-producing chemicals when it is manufactured. You would have to give up clothing, newspapers, food that is frozen, canned, or preserved in any way, and an almost endless list of other necessities. Oil, gas, coal, and atomic energy all add pollutants to the air. Living a pollution-free life is practically impossible, and few pollution fighters are aiming for this. They believe the problem occurs today because people were not alert enough or informed enough to look far ahead yesterday. They are working to remedy the problems and to plan for tomorrow to avoid additional ones.

Here is an example of a group of people who thought they were contributing much to fighting pollution. They met one afternoon and discussed the terrible state of the world. The leader of the group distributed several earthworms to each individual and, after a discussion of the benefits of earthworms to soil, the group left for their homes in highpowered cars that polluted the air. They turned up their radios and tossed a few beer cans from their cars on their way to their homes. Some were tired of the whole idea, while others continued their fight against pollution by introducting the earthworms to the soil of their gardens. Some put on buttons proclaiming them pollution fighters, while others felt that they had made their contribution and could go on with their usual way of life, except for an occasional outburst against industry and other contributors to the pollution of Planet Earth.

Many pollution fighters are well aware that the fight will be a long one. They consider the need for changing values, the need for education of every policymaker, professional, and citizen toward a better understanding of the world around him and his relation to it. True pollution fighters make new human values, attitudes, and behavior part of their lives in an effort to help solve this problem, which cuts across economic, social, and racial lines. In the country, in the suburbs, and in the cities pollution fighters are needed; for no matter where one lives on the Planet Earth, he is threatened by pollution. If a passenger on Planet Earth is not part of the solution, he is part of the problem. This is an issue which young people understand, and an issue on which many are acting. They are part of the spearhead of a challenging and exciting movement to protect the earth by preventing further pollution and attempting to restore it to a more natural balance so that it can support man and other life in future years.

2 PIECES OF A PUZZLE

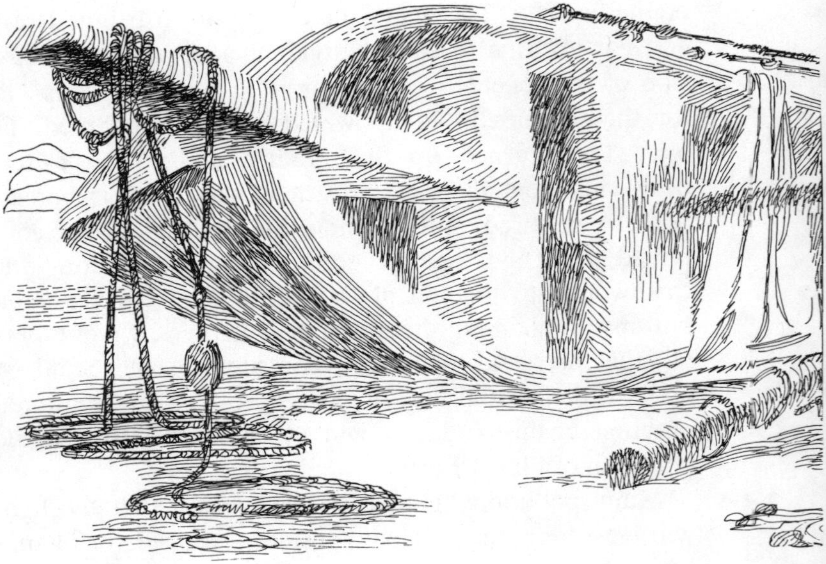

Pollution fighters have become aware that in nature it is impossible to do "just one thing." The "web of life," "cycles of nature," "delicate balance" are becoming popular expressions with people who are concerned about the environment. There is give and take in nature, on a small scale and a large.

Man, who once tried to control nature, has found that he fights a losing battle. He cannot win the struggle over animals, for he himself is one. He is part of nature.

In today's world, pollution fighters are well aware of their place in the natural scheme. But for many people, living as part of nature will mean a change in their way of life.

One of the most important tasks of pollution fighters is to make people young and old aware that the world

is a closed system, with limited supplies, and that when one changes a part of nature he sets in action many other changes. The use of materials for man's comfort and pleasure must take into consideration that there are many pieces to the puzzle of the balance of nature.

Past mistakes and upsets can be used to prevent similar damage. On this crowded planet, man must learn how some of the pieces fit together and how not to destroy any more of them.

Animal populations grow and decrease depending on the environment. This often happens in nature without the interference of man. For example, 5 million square miles of Arctic tundra, or one-tenth of the earth's land area, rings the North Pole with what appears to be barren land. Although it is treeless, the tundra is green, and is the home of countless living things. They range from simple plants such as lichens and mosses, to grasses, birds, foxes, caribou, and other complex forms of life. Lemmings, small mouse-like animals, live year-round. Periodically their numbers swell so tremendously that there is not enough space for all of them. Then they migrate in large numbers, stripping the land of food as they go. According to legend, these animals swim into the river to commit mass suicide. But scientists believe that the lemmings just continue to move forward in one direction. When they reach a wide river or lake, some animals cannot swim well enough to reach the other shore. If the lemmings come to an ocean, they will swim forward until they drown.

The number of foxes in an area depends to some extent upon the number of lemmings. In the years when there are large lemming populations, the foxes have a large food supply and they in turn can multiply rapidly. The number of caribou and the number of wolves are interdependent, too, for the wolves hunt caribou for food. This is just another example of the complex dependence of animals and plants in an unbelievable variety of processes and cycles in the tundra. So it is with living things, from the Arctic icecap to tropical rain forests, from arid deserts to lush grasslands.

Much as living things are interrelated, so the whole ecosystem is related one part to another. Sometimes man succeeds in changing one kind of environment to another. He irrigates the deserts, clears forests and creates grasslands, and fills in marshes to create dry land for building. In doing so he may upset the natural balance of things. For example, in their efforts to provide better waters for people, men find that they have upset the natural ecosystem so much that it may be many years before a balance can be restablished. The Aswan high dam in Egypt now controls the emptying of the great narrow river more than 600 miles upstream from the Mediterranean Sea. The great dam, which helps man to control the waters that flow to irrigate the crops in the farmlands below, seems to have created more problems than it solved. For example, since the downstream flow of water is slower, the sand does not build up along the delta, where dunes and sandbars once protected the fertile farmlands there from the invading sea. The millions of tons of rich soil that once were carried to the sea by the flooding Nile no longer flow with it, and the fish that used to feed on the animal life that was carried in the soil are disappearing from the river. Many of Egypt's fishing families have gone to dwell in the cities.

Another problem is in the spreading of worm-like animals which are parasites and carried by snails. These animals cause a disease that is commonly called snail fever. The animals that cause it infest the irrigation canals that are being dug below the dam. Although there is more water for crops and the green ribbon of fertile land that follows the Nile is becoming wider, the number of people exposed to disease from these infested waters is much greater than ever before.

Another problem exists in the lake created by the waters that are held back by the Aswan Dam. Wild water hyacinths are spreading over the surface. These plants evaporate water. Some people fear that the Nile may lose so much water by evaporation that the amount available for irrigation may be not much greater than before the dam was built. If the water hyacinths are poisoned, the lake water would be polluted and the water of the Nile flowing from it would be polluted too. Here are problems that have arisen because men still are not fully aware of the interrelationship of all parts of nature.

In one of his poems Francis Thompson tells us one cannot stir a flower without troubling a star. This may be poetic exaggeration, but it symbolizes many of the ways in which living things in the world are related. Take, for example, the story of the fish and the otters. Both fishermen and otters take fish from the water. It is not surprising to find that through the years fishermen believed that they could increase the fish supply by decreasing the number of otters. But such was not the case. Otters catch the fish that are weak with disease rather than the healthy fish. When the weak fish are not removed from the water, disease spreads among the healthier fish. As a result, there may be no fish at all for the fishermen. Similar problems occur when man upsets the balance of nature in many areas without fully understanding the web of life.

What do hippopotamuses have to do with an increase in snail fever? In South Africa, men waged a campaign against hippopotamuses. They were a nuisance in the rivers, so they were shot in great numbers. Before long, schistosomiasis increased among the people. What was the connection? The hippopotamuses kept the river silt in motion when they bathed. They also made natural irrigation channels when they climbed up the river banks, where they acted somewhat like bulldozers. When the hippos were gone, the rivers soon filled with silt, since there were no overflow channels, and periodic floods swept over the adjacent lands. This condition was favorable for the water snails which carry the fever. The people who shot the hippopotamuses, and their relatives and neighbors, suffered from an unwitting upset in the balance of nature.

The story of the prickly starfish, which is commonly

known as the crown-of-thorns, is another one in which man set off a plague because of lack of knowledge. This starfish has killed more than 90 percent of the coral along a 25-mile stretch of Guam's coastline in 2½ years, and has invaded 300 miles of Australia's Great Barrier Reef. Areas on the coastlines of other countries such as Fiji, Tahiti, Maylasia, Saipan, and New Guinea are being attacked.

Coral reefs are built from the calcium carbonate or lime skeletons of coral, which when alive are simple, stationary animals that attach themselves to the skeletons of their ancestors and build their lime homes on top of them. At night the coral unfold their tentacles and feed on small sea life, including millions of starfish in the larval stage. Today the starfish in the adult stage are eating the coral. No one knows exactly why.

Several ideas have been suggested to explain the problem of the starfish and the coral. In some areas, men destroy coral reefs by blasting and dredging channels, and this might have caused an increase in the starfish population. Another answer might be that too many giant tritons, a kind of mollusk which preys on starfish, were collected. The giant tritons have beautiful spiral shells and are sold for as much as $35. If these mollusks could be raised in great quantity in protected areas and returned to the sea, they might prey on starfish and help to control the present plague.

The destructive process in the coral reefs is the result of a chain reaction. Whether the population explosion of the spiny starfish is a natural periodic population growth or whether it is caused by one of the reasons suggested earlier, no one really knows. Pesticide residues which are washed into the ocean from the land have been suggested as another cause of this problem.

The story of the sea urchin's population explosion is somewhat better understood. Sea urchins feed on kelp, a form of seaweed. When the sea urchins have eaten all the kelp, their population decreases until more of this seaweed grows. Today, much sewage which contains living and dead plants is being washed into the sea. Sea-urchin population does not decrease for lack of food, for it eats the sewage supplied by man. The kelp is also eaten by the sea urchins; but since the animal population never decreases, the kelp does not have a chance to grow large.

Sometimes the trigger that upsets the delicate balance in an ecosystem can be easily identified. Such is the case in the rabbit problem in Australia. Here, a hunting enthusiast set two dozen wild rabbits loose on his property about a hundred years ago; the environment was ideal, and there were few enemies such as foxes to keep the population of rabbits under control. Before a hundred years had passed there were about five hundred million rabbits in Australia, many of which have been destroyed by the introduction of a virus that attacks only rabbits. This virus was introduced after a careful study to make certain it would not harm other creatures. Today, the killing power of the virus disease seems to be weakening, but at one time the virus killed as many as 99 rabbits out of 100.

The story of the cats in Borneo is another interesting example of man upsetting the delicate way of life. Here, DDT was used to get rid of large numbers of flies. Lizards of Borneo ate the flies and absorbed so much of the DDT that the cats who ate the lizards were poisoned. Many of the cats died as a result, and rats, which normally would have been eaten by the cat population, reproduced in large quantities and now plague the entire country. These animals are more of a nuisance than the flies who were originally the target of the program.

In March, 1970, mice swarmed across the richest grain and grazing country of Australia like a moving carpet, devouring the plants as they passed, and invading the homes, hospitals, and other buildings. Highways were reported to be so covered with mice that cars could not travel faster than twenty miles per hour because the mice made the roads slippery. Plagues of mice have been reported in northern California and Oregon, too. Some scientists believe that mouse poisoning by people is a major cause of the population explosion of the mice. Although the poisons kill some of the mice, the mouse population springs back to high levels more quickly than those of other animals which are their natural enemies, and which have also been poisoned by the chemicals intended for the mice. Man is blamed for the upset in the balance of nature.

What can you do to help prevent such upsets? Each individual has a small impact on the physical part of his environment and the environment has an impact on him. This is true of a tiny spider or a tremendous whale, of a delicate flower or a hardy weed, of a simple one-celled animal or an unbelievably complicated human being. Each plays his part in the many cycles of the environment.

Man can use his brain to look back on the many mistakes he has made in thousands of places in thousands of ways. He can look forward to more careful planning so that he will avoid further mistakes. He can act by taking steps to clean his spaceship so that it can better support the crowded passenger list of today. And he can promote awareness, learning, and action on many fronts.

You will find many suggestions in the following pages that will help you to be a better pollution fighter.

3 LIFE SUPPORT SYSTEMS FOR

Suppose you could tag an oxygen atom that is entering your nostrils right now. What path does it follow? It came from the air. It goes up your nose, down your trachea or windpipe, into your lungs. The oxygen atom, along with millions of others, travels into a small sac in the lungs where it passes through tissues and enters the bloodstream. It is combined with chemicals which, through complicated process, supply energy for your body. This particular oxygen atom may help to repair the muscle that was changed through use in skiing or swimming or running or just walking around the block and needs repair. It might help to provide your body with energy in changing the food you eat to skin cells that are replacing some that were sunburned off. The oxygen atom could do any one of millions of things in your body; but it does not disappear. It becomes part

of a tissue or part of waste matter that leaves your body.

Suppose this particular atom of oxygen has, through numerous complicated processes, become part of the liquid that is secreted by glands in your skin to help in the cooling process, or temperature control, of your body. The oxygen atom now is part of your perspiration. It leaves your body and goes back to the air. It is no longer free oxygen. It is now part of the water vapor in the air and it is combined chemically with hydrogen. Where does it go from here?

The water vapor that left your body could travel with the wind to different places. It could rise into the sky above and become part of a cloud that drops its rain before it blows far from where you live. Suppose the tagged oxygen that you breathed long ago is now part of water that falls with millions of other atoms onto the ground in the form of rain. And where does it go from here?

The water sinks deeper and deeper, for there is much rain and the soil does not hold all of it. The tagged oxygen finds its way into the water table, which lies below the soil. From here the tagged oxygen you once breathed may travel, with other oxygen and hydrogen atoms that make up water molecules, for a very long distance or a very short one. Suppose it reaches a well in the country not far from where you live. A pump carries the water into a house and when someone turns on the faucet the tagged oxygen flows out as part of the drinking water for another person.

The tagged oxygen that was once in your body may come to the surface of the earth again in the water of a spring, which may feed a stream that eventually goes into a river. From there it may begin its long journey to the sea. There is an almost endless variety of things that could happen to the oxygen which you breathe. No matter what became of it in your body, no matter how long it stayed there or how it left, no matter where it went after it left your body, you can be sure of one thing: the oxygen was not destroyed. The oxygen eventually would be used by some other living thing.

Much of the oxygen you breathe in is breathed out again in the same form. This is a part of an oxygen molecule (O_2). Some of the oxygen which you breathe in is eventually breathed out as part of the well-known chemical, carbon dioxide. Much happens to the oxygen, chemically, before it combines with carbon and is breathed out as a gas which becomes part of the air. Perhaps you are wondering if there will be enough fresh oxygen when the earth becomes increasingly crowded with people. The trash, or garbage in the sky, or the chemicals men have added to the once-fresh air, will be discussed in a later chapter. The way in which the supply of oxygen is renewed is the same way it has been renewed for millions of years on Planet Earth. But the amount may decrease if the balance of life is upset too much.

The green earth has long produced fresh oxygen, which normally makes up 20 percent of the air that surrounds the earth. The carbon dioxide which is breathed out by the earth's animals and to a lesser extent by all plants, is used by green plants, which make food when the proper amount of light is present. They do this through a complicated process, photosynthesis, which man is still trying to completely understand. But what goes in and what comes out fit simply and nicely with the needs of animals, forming the famous oxygen-carbon dioxide cycle. So it is that the supply of free oxygen in the air is replenished through a balance established through millions of years: Carbon dioxide and water in the presence of light and chlorophyll produce oxygen and food.

Although relatively small amounts of oxygen are combined with other chemicals, due to rusting and other processes, the changes which take place on the planet keep the oxygen supply much the same.

Water too is part of a seemingly simple cycle. The water in the streams, rivers, ponds, lakes, and oceans again and again rises through the air in the form of vapor, gathers in clouds, and drops back to earth in the form of rain. Here the water is used again and again. So it is with the water that you drink; but the process is far more complex.

If you could tag a water molecule that is one of the billions in a single glass of water that you drink, where might you find it if you had a way of following it? From waste water to drinking water is often a long and complicated cycle; but unfortunately it is not always long or complicated enough for the water to naturally cleanse itself. To complicate matters, pollution from a wide variety of sources makes waste-water treatment difficult; and while the amount of rain remains about the same each year, the world's population grows, and industry and agriculture use increasing amounts.

From the earth to the sky and the sky to the earth is one kind of water cycle. From pure, clean water to waste water, back to pure, clean water again, is still another kind of water cycle.

Even the land part of the earth itself is part of a cycle. The soil that forms a thin layer on much of the land part of the earth was once rock. Through long years of grinding, wearing away by winds and sand, or the action of animals and plants that lived on the rock, smaller and smaller pieces of rock were changed to soil. Soil is also composed of organic matter, decomposed animals and plants.

The making of just one inch of soil may take as long as thousands of years. Since the plants which supply man's food depend on the soil, it is no wonder that man has long been trying to conserve the supply. Growing things may deplete the quality and the supply, and water washes much of the soil toward the sea. The large numbers of animals, especially the human animal, are interfering with the soil and rock cycle, just as they are with the oxygen-carbon dioxide cycle and with the water cycles of the world.

Another part of the Planet Earth's cycles is one in which animals eat plants and, in some cases, plants and animals. Both of these contain carbon, which they use for their own needs. When plants and animals die, their bodies contain carbon in compounds, which eventually is released as carbon dioxide when their remains decompose in the soil after their death.

Another portion of the carbon-oxygen cycle releases carbon dioxide from plants and animals that have been part of the earth for thousands upon thousands of years. These are the plants and animals that made fossil fuels which we burn today. Coal, gas, and oil were once living things that breathed oxygen. When their remains are burned, carbon dioxide is released in the air.

The nitrogen cycle is another way of natural recycling on Planet Earth. Green plants absorb minerals from the soil. Some of these are nitrates, compounds that contain nitrogen, which is chemically combined with oxygen and usually with sodium or potassium. Plants use nitrates along with other chemicals to form amino acids, the building blocks of proteins. Proteins are used by the plants for growing.

Animals also form proteins, and they get the amino acids when they eat plants or other animals. But not all proteins are used in the building of plant and animal tissue. Some combine with oxygen to produce energy. The waste products are nitrogen compounds such as urea, which are excreted and return to the soil or water systems of the earth. Here bacteria decompose the compounds and produce ammonia, which is a combination of hydrogen and nitrogen. Other kinds of bacteria in the soil combine ammonia and oxygen to form compounds known as nitrites. These compounds cannot be used by the plant, for they are not absorbed by the plant roots. But other bacteria combine chemicals to make nitrates, a form of mineral compound that can be used by green plants. The cycle here begins with nitrates and ends with nitrates.

Although green plants cannot build proteins with the

pure nitrogen that makes up about 78 percent of the air, this nitrogen is involved in the nitrogen cycle. Two groups of bacteria combine nitrogen with oxygen to make chemicals known as nitrites and nitrates. One of the groups is a soil bacteria, and the other lives on the roots of plants known as legumes. Clover and alfalfa are examples of legumes, which farmers often plow back into the soil because of the beneficial bacteria that grow in their roots.

Another element in the nitrogen cycle is the conversion by lightning of small amounts of nitrogen from the air into nitrates. Every flash causes some nitrogen in the air to combine with oxygen, forming compounds. Such nitrates are carried to the soil by rain or snow.

Year after year, century after century, the cycles of nature continue. In addition to the cycles of their physical environments, the tremendously complicated food chain described in a later chapter supports a balance of life on Planet Earth. The balance fluctuates, depending on the number of natural enemies and the conditions that favor certain animals and plants at certain times and places. But the complex interrelationship between plants and animals, and between prey and predator, exists in all kinds of ecosystems.

The animals and plants of Planet Earth are related in such a complex way that we cannot grasp the web as a whole. Cycles are repeated on a small scale and on a grand one. All of these factors are somewhat like actors in a drama of an ever-changing environment.

Of all the passengers on Spaceship Earth who must depend upon a closed system where supplies are limited and recycling means survival, the villain is the thinking animal, man. Because man has tried to tame nature, he has not recognized that he is a part of it. He may be the star, but the play will close if the supporting actors, the other animals and plants, are destroyed. Man must learn to recycle, or use and reuse, on a larger and bigger scale to prevent this.

4 HOW TO HELP SAVE A POND

Have you ever thought about a lake as having a time of birth, of being young, maturing, and growing old? This is a natural process called eutrophication. Lakes do not live forever, but some live longer than others. One of the factors that can make a lake grow old very fast is its treatment by man.

You can help to save a pond or lake from premature death in a number of ways. But first you must know something about its life story.

Water exists underground in all parts of the earth, even beneath the Sahara Desert, and most of it is constantly in motion. Some water emerges to the surface of the soil in the form of springs. Some water seeps into streams, rivers, and lakes through underground springs.

Ground water from springs and rain water together make lake water. The amount of water in each lake

OR LAKE

varies, partly because of the changing level of the ground water and partly because of the difference in the amount of rain water that falls from time to time, the temperature, and other weather conditions.

You can form a tiny lake in a bowl. Put some loose soil into the bowl so that its surface is about three inches below the top. Gently pour water at the edge of the bowl so that it can sink into the soil. When the soil is holding all the water that it can, dig a small hole in the center of the soil. Watch water collect in the hole. Here is your pond or tiny lake. So it is with nature; for when the ground is saturated and the land is low, water lies in puddles, in ponds, and in lakes.

Many factors play a part in how long a pond or lake lives. All the streams that enter a lake carry dirt into it. Streams leaving a lake cut back into it and speed

convert the dead algae into plant and animal food fast enough to maintain the natural balance that once existed. Generation after generation of algae fall to the bottom and add to the layer of muck. The sediment grows thicker and the water more shallow, while at the same time the plants at the edges of the water make the lake's boundaries smaller. So it is that overfed, or man-fertilized, lakes die a hundred or more times faster than most lakes that can maintain a natural balance.

Recently, there has been much concern about the heat that is being discharged into rivers and lakes from power plants, especially those which use nuclear energy as fuel. Studies are being made to determine the upper limit of water temperature that is safe for life in lakes and rivers. Thermal pollution is one of many types of pollution that needs much further investigation.

Is there an ailing lake in your community? What can you do to help to prevent it from becoming a "dead sea?" As an individual you can help to organize groups to save local lakes and ponds. Perhaps you can persuade community officials to harvest weeds and fish to remove the excess food from lakes. Weed-harvesting techniques have been used in a number of places. In some lakes, cropping one-third of the lake three times, to a depth of five feet below the surface, seems to be helping a great deal. Contact your local conservation society and some of the organizations listed on page 129.

In some communities, such as those near Lake Tahoe which borders Nevada and California, pollution fighters have banded together and have helped to change laws. Now wastes must be rerouted so that they do not go into the lake.

Although many people are aware of the foul odors and the unpleasant sight of dying lakes, they do not concern themselves enough to look for the causes. You can help to pinpoint the sources of nutrients that encourage excess algae growth and reproduction. Nitrates and phosphates are two groups of chemicals that are largely responsible. Some may come from municipal sewage-treatment plants. Some may come from runoffs from manured fields. Sometimes ground water contributes, along with industrial wastes and seepage from cesspools. Detergents from household and institutional dishwashers and home and commercial laundries add to the problem for they add as much as 40 percent of the extra phosphates. Do you use detergents with low phosphates content in your house?

Pollution is being controlled in some places where manure from cattle is stored in vats during the winter until the spring thaw, and then it is sprayed on fields at a time when they can better absorb it. In this way the manure does not run off the land into the lake. This system is expensive for farmers, for they must provide costly storage tanks and spraying equipment; but the amount of water and soil conservation may encourage communities to defray the costs.

Scientists are experimenting in many ways to reclaim ponds and lakes and rivers from the development of algae to the degree known as algal blooms. In addition to regulation of plant nutrients such as nitrates and phosphates, some scientists are using viruses that destroy some of the algae and control its population on the surface of the water.

The algae that create a problem in one lake may be very different from those that create a problem in another. There are many different kinds of algae. Algae in water samples from one lake produce a greenish-brown color when they are far less plentiful than algae samples of a different kind that barely show in another lake, even though they are ten times as plentiful.

One approach to helping an ailing lake is to encourage the growth of bacteria that compete with the algae for the nutrient materials that are provided by dying plants and animals. In a lake where there is a balance, self-cleansing is brought about by such bacteria. This is a natural process. Pollution fighters can make use of knowledge of how major cycles take place to reestablish balance and return the lake to its natural state.

Many say that Lake Erie is dead and beyond recovery. No longer does it support great commercial fishing industries. Sports on Lake Erie have been spoiled by pollution. But other people believe that Lake Erie is far from dead and can be cleaned up by removing severe pollution at the sources of chemical waste along the shore. This lake is surrounded by immense populations in the United States and Canada. Here, the shortened life span of a lake due to man's interference is considered a classic case. Although it may never be completely restored, it certainly might be improved by community action.

Changing the fish population of a lake is one way which some scientists hope may improve its quality. For example, many of the fish that live in an aging and smelly lake are not the kind that fishermen seek. Sometimes overfishing has encouraged the growth of animals such as lampreys and alewives. If desirable fish were added to an aging lake, the food chain would be put on a better basis. A large salmon could eat many alewives and other undesirable animals and plants in the biological food chain. Salmon could be removed from the lake by sportsmen and commercial fishermen. More must be learned about what kinds of desirable fish can live with low supplies of oxygen and in lakes where there are temperature problems. This is one of many ideas that need further research.

If you live near a lake, collect a jar of water from it. Even a drop of water is teeming with life, but most of the plants and animals are too small to be seen without a microscope. If you observe some of the water in your jar, you may find various forms of algae, small animals known as protozoans, insects, small worms, crustaceans (crayfish, shrimp, daphnia), and other living things. Perhaps someone can help you to stain the living things in the water so that you can see the microscopic ones better.

The teeming life in a lake is extremely varied and each kind participates in the cycle of life. There is much to be learned about the life story of lakes and how man can change the situation from one in which lakes are dying faster than new ones are being born. You can help, as a pollution fighter, to encourage more research on the causes and remedies for ailing lakes. The warnings are obvious; the cures may be many. But fewer lakes mean less drinking water, less food, and less beauty. The challenge is worth the effort.

5 WETLANDS OR WASTELANDS IN

Perhaps the least understood ecosystems of the Planet Earth are the wetlands, which include the bogs, swamps, marshes, and other coastal regions. Although they may appear useless and are sometimes unattractive, their value is great. There are salt water marshes and fresh water marshes. There are deep marshes and there are shallow ones. There are shrub swamps and wooded swamps. There are marshes that are regularly flooded and marshes that are irregularly flooded. There are bogs that support a spongy cover of mosses and there are bogs that support plants such as pond pine, sweet bay, and pitcher plants. All of these wetlands play an extremely important part in the balance of the fragile planet on which man is but one of several million kinds of dwellers.

If you happen to have a wetland near you, perhaps

YOUR FUTURE

you are wondering whether it should be called a *bog*, or a *marsh*, or an *estuary*. Although these are related, they differ in the following ways: A *bog* is a form of wetland that has large deposits of peat, with floating mats of sphagnum and shrubs such as cranberry. Sedges are common in bogs. These are plants which are similar to grass, but have solid triangular stems rather than the round hollow stems of grasses. Decayed remains of sphagnum moss plants make up the major part of the peat. One often finds spruce and cedar trees in bogs. This type of wetland usually develops when a fairly deep lake with poor drainage is dying.

You may live near a wetland that is treeless and is a *true marsh*, which developed in an aging shallow pond or river margin. Many marshes are parts of *estuaries*, places where the rivers and oceans of the world meet.

Some of these areas extend for many, many miles and are extremely important in the fragile balance of Spaceship Earth. Here the fresh water and the salt water mix. Here marshes and mud flats are sometimes completely dry and sometimes drowned by salt or fresh water. The tremendous circulation of water supplies a rich nutrient for the many plants and animals that live at various levels of the estuaries. There are land plants with roots in the mud and ocean plants that drift in from the sea. Many fish come to the estuaries to lay their eggs on the pebbles or grasses that grow under the waters. The larvae of the shellfish are nourished here, where plants feed the animals that come from the sea. Birds search for food as the tide flows out. Such wetlands thrive with an almost endless number of creatures and a wide variety of plants.

Sometimes marshes in shallow water basins might become *swamplands*. *Swamps* also develop in sluggish streams and in places where the water floods again and again until some remains. In a swamp, there are trees such as cypresses and gums and there are shrubs and mosses. Swamps differ from bogs in that they usually have better drainage.

Swamps, marshes, bogs, salt-water wetlands, freshwater wetlands, or wetlands that combine salt water and fresh or brackish water all have their own typical plants and animals and distinct ecosystems. The plants and animals in each one are interrelated, just as each kind of wetland is interrelated to the plants and animals of the rest of the world.

Those who think of wetlands as wastelands have missed much. If you can explore a wetland of any kind, you will find just for the looking some beautiful plants and animals in an environment that is often considered

unpleasant. Use a book such as William A. Niering's THE LIFE OF THE MARSH, McGraw-Hill Book Company, as a guide for exploring wetlands.

Even if you personally never can wade through the marshes or bogs, or explore an estuary for worms, crabs, oysters, shrimp, clams, wrens, ducks, geese, owls, terns, grasshoppers, muskrats, minks, rabbits, possums, raccoons, crickets, ants, wide varieties of fish, and the numerous other animals that live there or come there to feed or reproduce, you can join those who are fighting to perserve wetlands from the invasion of man.

The wetlands are havens for countless living things, and they have developed over millions of years. Many people do not realize that they are the nurseries for many of the fish man uses for food. As a pollution fighter you can do much to help preserve this important part of Planet Earth by making people aware of the value of wetlands and the ease with which they can be destroyed.

Men find what they believe are better uses for marshes than that of maintaining a balance of nature. In some areas, the wetlands are drained, filled in, or mined. Sometimes channels are dredged. In other areas the wetlands are used as sewers, until they are smothered with garbage and trash. The once-fresh salt air that blew in from the edge of the sea across the marshes smells like rotten eggs in many of the areas, for man is destroying millions of acres of wetlands.

Point out to those who say such lands are useless that scientists find from experiments that estuaries are twenty times as productive as the open sea and twice as productive as a cornfield. The birds they support fly to areas where they destroy insect pests. Where marsh grass is replaced by some algae, valuable shellfish and other foods that support the fishing industry no longer breed. Channels are deepened for boating; but the stirring up of silt by dredges makes life more difficult or even impossible for animals other than men in boats. The world of nature is soon replaced by a world of oily water, scum, and scavenging sea gulls.

Even the tidal pools that provide homes for rock-bound marine life are suffering from the actions of men. Although in some areas collecting is illegal, many tidal pools are still being invaded by collectors who remove large numbers of sea urchins, starfish, octopuses, crabs, and other specimens. The tragedy is that the collectors allow these salt-water species to die because they do not have the proper aquaria for housing them.

On Planet Earth, marshlands both large and small are becoming the concern of communities. For example, on Long Island several hundred residents gathered one

Sunday to demonstrate against the building of a parking lot in the salt marsh of Udall's Cove. The group included about 350 children, students, and adults who were concerned with the loss of one of the last natural wild areas in or near this region. By studying why the marsh grasses were important, and by explaining to their neighbors about their part in the food chain, people were aroused through a wider area, and politicians became involved. In such ways small groups of individuals can make their interest felt. This is ecology in action.

The saving of the Everglades is an example of ecology in action on a much larger scale. This unique park has been called Florida's River of Grass, for here there is a belt of tall grass that runs for about 95 miles and is as wide as 40 miles in some places. Once this subtropical national park is lost, it can never be replaced. Some look to it as a warning system for all the life of Southern Florida. Under the leadership of David Brower and the National Audubon Society, people brought attention to the problem of survival of the Everglades and its importance to the water supply of Florida. When the jetport which was being built in the Everglades was two-thirds finished and about 13 million dollars had been spent, they were successful in obtaining legislative action to preserve this watershed area for future generations and to preserve a natural filtration and storage system of water so vast and so complex that man could not duplicate it in years to come. In addition to its function in water storage, the Everglades provides a home for almost countless plants and animals. Alligators keep large water holes open so that when the dry season comes these wet areas are available to many animals that need wetlands. Egrets, buzzards, soil grass waving in the wind as far as the eye can see, sunlight on Spanish moss are indescribably beautiful. Unless one visits the Everglades and explores its quiet and strange beauty, he can not really appreciate the importance of this part of nature. Continual action is necessary to preserve it for the future. Pollution fighters must help people to prevent its further destruction.

Here is another example of pollution fighters at work. San Francisco Bay was originally 700 miles square. Developers in the past hundred years have filled all but 420 square miles of it. The Sierra Club and other conservation groups organized a Save Our Bay movement, which resulted in government and state action to curtail further destruction. The Bay Conservation and Development Commission, which was established in 1965, was continued by new legislation in 1969. Concerned citizens are thus helping to protect thousands of acres of tidelands and marshlands, and to prevent the serious effects of more filling of the Bay. Not only was there danger of reducing the number of fishes and other animals that lived there and destroying the beauty of the region, there was also concern about severe changes in climate in nearby areas if the Bay was not saved.

As a pollution fighter, one of your most important tasks in the preservation of Spaceship Earth is arousing awareness of the value of such misunderstood areas as the wetlands. Just as the crisis in the Everglades has been brought to the attention of many individuals and action is still needed to preserve some parts of this tremendous sea of grass, many areas need continued help. The delicate balance must be kept intact by constant vigil of the once-lush wetlands that are becoming parched and dry. You and other pollution fighters must guard the bogs, the swamps, the marshes of the world and help others to appreciate the value of these strange and fascinating ecosystems.

6 RIVERS AND SEWAGE TO THE SEA

What do you find when you cruise down today's rivers? No one expects every river to be sparkling and clear as a swimming pool, although the Amazon River and many mountain streams are still quite pure. Such sterile water is not natural and rivers never were completely pure. For years on end they have carried sediment and a variety of chemicals and living things as they traveled to the sea. Boat owners are becoming more conscious of the fight for clear water, for it affects them in obvious ways. Each boat owner can play a part in the vast new war on water pollution. The following are not only good practices to help prevent pollution, but they are good manners:

1. Don't throw trash into the water.
2. Be sure there is a trash receptacle aboard your boat.
3. Deposit trash in the proper container at the dock.
4. Insist that docking areas be equipped to remove waste from boats.
5. If your boat has a head (toilet), be sure there is a holding tank or adequate means of waste treatment aboard.
6. Support local, state, and federal water pollution control programs. For greater protection and convenience, help promote uniform state boating laws.

Although waste from one boat may seem very unimportant, the waste from many boats contributes quite a bit to the water pollution problem. About 10 million pleasure boats are using streams and lakes in the United States. If each of the million people who enjoy boating joins the fight for clean water, the boating in the future will mean more pleasure for them and cleaner waters for everyone to drink and enjoy.

Suppose for a moment that your drinking water comes from a river into which many streams empty. Even though a fair amount of waste material is dumped into the river, it is self-cleansing as it travels along, and by the time it reaches the place where your drinking water is drawn, the river may be quite pure again. Unfortunately, with so many people on today's Spaceship Earth, the amounts of sewage, garbage, fertilizers, mine drainage, industrial waste, and sediment that pour into rivers make it impossible for many rivers to renew themselves. There are numerous places where rivers serve as both sewers and a source of water supply. Many of these rivers are so contaminated that even man's best systems for purifying drinking water are unable to do a good job on their water.

Have you noticed an increase in the taste of chlorine in your water? In about half the communities supplied in the United States, only chlorine is used to kill bacteria. You drink the pesticides, herbicides, and other chemicals in the water, if you live in such an area. Every year more than 500 new or slightly different chemicals are introduced to the United States, and many of these ultimately find their way into the rivers that supply drinking water. Studies must be made to find out the long-range health problems that may be brought about by these chemicals.

What do you know about the purity of the water that comes from your faucet? Does your community have a waste treatment plant? How does it function? Do industries dump their untreated waste into the water that is used eventually for your community's drinking?

Consult your local health department to find out more about what is happening to your drinking water. Find out what happens to the sewage your city or town produces. Does it eventually get into someone else's drinking water? Can something be done to prevent it from contaminating the water in rivers that run to the sea?

Suppose you take a trip in a boat that is inspecting an industrial river. Some people say that an industrial river is actually an open sewer where wastes from various industries are poured. At one point your boat runs aground on the muck that lies on the bottom of the river. Later, as the boat chugs along, you notice unpleasant odors. In the summertime the odors are far worse than they are today, for then the heat hastens the decomposition of the sludge on the bottom and this uses up more of the oxygen. The river bubbles with waste gases, releasing even fouler odors. Fortunately, today the wind is blowing and the air is cool, but you do notice airborne odors from industries.

Now a scoop is lowered to bring up the samples from the bottom. You see a black gelatinous mass that smells like oil. As you look around, you see soda cans and beer cans floating in the river and on the banks of the river.

If pollution fighters could do enough, might such rivers be changed from grimy, oily places to pleasant ones where green grass grows along the banks and people can swim and fish. Such action would take a great deal of money, but there are many people working to help purify the water that flows to the sea. Perhaps you can help to reclaim a river near your home.

There are many ways to help save a river. In the summer of 1969, folk singer Pete Seeger sailed the Hudson in a sloop which was a full-sized replica of the original Hudson sailing sloops to make people aware of pollution problems along this river. The sloop *Clearwater* docked in many communities and served as a floating museum that was visited by large numbers of school groups and interested people.

Back in the days when the Hudson River sloops sailed in the clear waters of this beautiful river, it was safe for swimming, and many fish were caught for sport and by commercial fishermen. Today the refuse from hundreds of towns and factories has dimmed the once-clear water. In some places dead fish frequently wash up on the beach. At the town of Castleton-on-Hudson, New York, fish washed up on the riverbank six inches deep, and their unpleasant odor permeated the whole community. The fish had to be buried by bulldozers. At Castleton, the Hudson carries the murky brown waste from a number of cities north of it. The oxygen supply is so low that at times even the scavenger eels cannot live there.

As it flows toward the ocean, there are places where the majestic Hudson manages to cleanse itself to a fair degree. But as it pours past Manhattan and into the bay and the ocean beyond, conditions are so bad that this area has been called a "dead sea."

The "dead sea" area near the entrance to New York Harbor covers a space of twenty square miles. All the waters within a six-mile radius of the contaminated area have been closed to shellfish harvesting. Fish captured in this area suffer from a disease which may cause their tails to disintegrate. Lobster gills have been found covered with spots of oil and debris. Shellfish harbor hepatitis, which can be transferred to humans who eat them. Here as in many other areas, pollutants have exhausted the oxygen in the water and prevented light from penetrating to the plant life beneath the surface, thus stopping photosynthesis. The condition is so bad in the New York area that even some of the animals such as scavenger eels that tolerate pollution conditions in other areas cannot live in some of these waters.

The Ohio River has been polluted for more than twenty years. This 981-mile long river was once known to French explorers as La Belle Rivière, or The Beautiful River. Today you can see pipes from a plant that combines steel with other metals spewing bright green, poisonous waste fluid into the Ohio's water. At another place along the river, iron oxide stains the water a reddish brown when the wastes from the steel blast furnaces are poured into it. In still other places, one finds some bluish black goo and splotches of bright green and blue that add both color and pollution to the once-clear and sparkling waters. In some places along the river, human sewage is dumped into it. At others, partially treated sewage flows lazily along and mixes with the river water. The once-beautiful Ohio is a broad and placid river still; but its load of chemicals, oil, grease, and other pollutants make it unpleasant to look at and to smell. Its waters are unfit for swimming, and fish life has practically disappeared.

Men are continuing their efforts to clean the dirty Ohio so that the dead fish and raw sewage may someday be replaced by healthy, live fish and the kinds of plants and animals that live in unpolluted waters. These living things, both large and small, help to cleanse the river when they are not overcome by too much sewage.

People in the Cleveland, Ohio area used to joke about the nearby Cuyahoga River being a fire hazard. Industrial wastes in the river have long made the waters oily. Not long ago, the river really did catch fire, causing thousands of dollars' worth of damage to two railroad bridges. Actually the blaze started on a large oil slick on what is sometimes believed to be the most polluted river in the world. The State of Ohio is trying to enforce its antipollution laws against communities and industries that discharge oily waste into this river that flows into Lake Erie. But much work and many years will pass before the Cuyahoga pours clean water into the lake.

What is happening to the streams and rivers near your home? On a map, follow the paths of these rivers to find where they empty their water. Do they eventually reach the ocean? It has been estimated that the seas of the earth contain 330 million cubic miles of water. This is 18 times as great as the volume of all the land above sea level. No wonder man has always considered the oceans as dumps that could absorb endless amounts of material.

Long before the time of man, chemicals from the land had been dissolved in the rivers and carried to the seas. The list of elements found in the sea is long, but the amount and the number is changing as man invades it with his chemicals. It has been estimated that man has been influencing the chemical makeup of sea water more than any of the animals or plants that live within the sea. Nearly all the materials used by man are eventually discharged as waste, and a major portion of this discharge finds its way to the sea. Some comes from rivers; some falls into the ocean from the air above; and some is spilled by vessels that travel through the sea.

Thor Heyerdahl, who crossed the Atlantic Ocean in 1970 in a craft built like those of the ancient Egyptians, had an excellent chance to look at the surface of the ocean. During the expedition, the ocean surface was just a foot or two below the edge of the vessel, and in many cases the deck was partially submerged in the water. Mr. Heyerdahl and his crew reported an alarming amount of pollution far from the shores. Even in mid-ocean they noticed lumps of black, tar-like material and cellulose from the remains of toilet paper drifting far at sea. At one point they reported that the water appeared dirty even though far from any harbor that emptied sewage directly onto the ocean.

From this report, and from the reports of numerous scientists, pollution fighters are becoming alarmed at the possibility that the oceans, too, may become so polluted that they cannot cleanse themselves and that the seas will "die."

From time to time, hundreds of birds are washed ashore along the beaches, their feathers covered with oil. At various beaches volunteers by the thousands join in trying to rescue such birds and return them to the sea by removing the thick oil that clogs their wings. After washing the oil covered birds with salad oil or detergents, the volunteers protect them until the natural oils return so that the birds will not die of exposure.

If you live along the ocean's edge and have seen birds covered with oil, you know how people working together can help to preserve the beauty of the beaches and save the birds. "Don't let the ducks die" was a movement in St. Petersburg, Florida, which involved the rescue of about 7,000 birds. There was a "wash-in" at one high school's biology laboratory. Fifth graders from another school chopped up fish that was bought at the store to help feed the hungry birds. Students from a college mixed a menu of lettuce, shrimp, and crackers. Boat owners worked night and day bringing floundering birds to the beaches, where volunteers worked thousands of hours to massage and clean these ducks and other waterfowl. Through the long pollution fight, Boy Scouts hauled wood that was donated by a lumber company and fed the fires under garbage cans to provide a constant supply of warm wash water for the birds. Young and old worked side by side in a scrub-and-brush project to fight pollution and save the threatened animals from the black goo that held feathers and wings in a deadly grip.

The people of St. Petersburg advertised in newspapers to send a message to the people of the world:

> *Beware! It's so easy to let the things we value most die step by gradual step. We had our dying ducks and threatened beaches to wake us up. And you? A small child said, "Don't die, duckie, don't die." Because if you do, so will we all, a little. Not just here in St. Petersburg, but wherever in America pollution threatens.*

One can see the ducks dying at the edges of the water. But what is happening far out where Thor Heyerdahl spotted pollution? The chemicals that man adds to the sea spread throughout the oceans of the world; for the sea is a great mixing bowl. Each ocean is connected to the other. If one were to dye red the molecules in a single glass of water, these molecules would spread throughout the world. If, thousands of years later, you could dip a glass into any part of any sea, some of these red molecules would be collected.

Dr. Sidney I. Holt of the United Nations Food and Agricultural Organization expressed an opinion that pollutants of the seas are increasing almost faster than man's ability to get information about them. In spite of the vastness of the seas, people are questioning more and more the effect of new chemicals which are being added to the oceans by the activities of men.

Mercury is normally added to the sea by way of the rivers, which wash it from the continents at a rate of about 5,000 tons per year. Another 4,000 to 5,000 tons of mercury enter the oceans as a result of man's activities; for mercury compounds are used in fungicides, in the treatment of seeds, as parts of processes which produce chemicals such as chlorine. As a result, fish and other sea animals accumulate in their bodies mercury compounds in much higher concentration than they would under natural conditions.

After a serious catastrophe in Japan some years ago, some scientific detective work pointed to mercury poisoning as the result of eating fish contaminated with mercury. One hundred and eleven cases of mercury poisoning in people were reported. Forty-one deaths occurred, and many people were paralyzed. Cats and rats in the region also suffered from eating contaminated fish. The source of the excess mercury was traced to a plastics manufacturer, who had dumped waste sludges into Minimata Bay, Japan. Fish that caused this tragedy were found to contain fifty parts per million of mercury.

Recently a graduate student in Canada discovered fish containing 1.3 to 7 parts per million of mercury in Lake St. Clair. The safe standard is .5 parts per million.

After this, fish on the American side of Lake St. Clair, Lake Erie, and other lakes have been tested for mercury and have been found to be contaminated. As a result, commercial fishing was banned in areas where these fish were found, and investigations are being made to find the sources of contamination.

More and more pollution fighters are realizing that, in spite of the size of the ocean, it is not an unlimited receptacle for trash, chemicals, and sewage. It cannot continue to be self-cleansing when pollution reaches large amounts. Oil spills at sea have been causing increasing concern to people who are aware of ocean pollution.

About 1,000,000,000 tons of petroleum travel across the oceans of the world each year. Only about one-tenth of one percent (0.1 percent) is lost through spillage and leakage; but this amounts to many, many tons of oil in the marine environment. Even some fish that reach the dinner table taste of petroleum.

Although the sea itself produces materials similar to the ones that make up petroleum or oil, the amount added by man disturbs the balance of the sea. While pollution fighters attempt to help in times of local disaster, when oil spills over beaches and cover the feathers of birds, other pollution fighters are working to try to prevent further chemical invasion of the ocean.

Solid wastes from sewers are discharged into the sea at the rate of about 500,000 tons per year from the 30 million people who live along the seacoast of the United States alone. On a world-wide basis, scientists estimate several million tons of solid sewage are contributed to the world's oceans. Although this amount of solid material is small when compared to the amount poured into the oceans by way of the rivers each year from the crusts of the earth, natural communities are altered at the areas where sewers dump their loads.

Radioactivity in the oceans has long been a concern of pollution fighters, for these vast bodies of water seemed a natural dumping place for the discharges from nuclear reactors and other waste products of atomic energy processes.

Rivers and oceans have been used as sewers since the beginning of time. Certainly the quantity of water dilutes the materials which are dumped into it. Certainly the problem would not be so great if it were not for the fact that some living things concentrate the poisons and radioactivity which are introduced into the water.

For example, DDT residues are passed along from sea plants to plant-eating animals, to animals that eat small animals in the sea. Then larger animals eat these animals, and so on. Birds pluck fish from the sea and people eat fish and birds. Each time living material is transferred from one link in the food chain to another, more than half of the DDT is excreted or lost through respiration. But the remainder becomes part of the living tissue. As the living matter is transferred to the higher link, the concentration of DDT increases.

Radioactivity increases in much the same way. Through food chains and food webs, yesterday's trash becomes tomorrow's food. A poisoned sea means a poisoned planet. What are you doing to prevent rivers from polluting the seas?

7 KNEE-DEEP IN WASTES

Can you imagine man standing on the earth knee-deep in garbage and trash, shooting rockets into the sky which will carry men to a lunar laboratory? Getting rid of trash and garbage in the orbiting laboratory is somewhat simpler than getting rid of trash and garbage on earth because of the enormous quantity of the latter. In the oribiting laboratory, materials can be recycled and used again. Changing trash and garbage into forms that can be put to use is known as recycling. Recycling of trash and garbage is gaining in popularity on Planet Earth; but relatively little is done because of the cost.

Some recycling programs are already going into effect on an experimental scale. Perhaps you live in an area where you can take aluminum beer or soda cans to a reclamation center and be paid half a cent per can.

The Reynolds Aluminum Company has been experimenting with this kind of antilitter aluminum-can recycling for several years, and is now putting such projects into effect in sixteen states. In Los Angeles alone, more than a million cans a month were brought to the plant. People were not only rewarded financially, but also felt that they were able to help to clean the streets, the beaches, and other areas, and to help in the reusing of a natural resource. Scrap aluminum is worth $200 a ton. Because it can be melted down and reused so readily, scrap aluminum is more valuable than many other kinds of solid wastes.

Scrap steel is recycled to some extent, and some steel producers are now making a chrome and resin-coated steel can that is more acceptable to the steel mills that use scrap.

Some glass companies are experimenting with the reuse of no-return bottles, which they buy back at about one cent a pound. The old glass is crushed into a material called "cullet" which is used in the manufacture of new glass containers. Scrap glass can be used in many new ways. A very small amount is being treated so that it can be used as part of floor tiles, insulation, and other construction materials. One interesting project is the development of a glass material which decomposes upon contact with water. Professor Samuel Holbert of Clemson College, in cooperation with the United States Bureau of Solid Wastes, has developed this kind of glass material. He also has developed various kinds of coatings which protect glass from moisture until the coating is cracked. Then most of the material decomposes into silica and other minerals.

Suppose you could run water over your soda bottle and flush it down the sink, instead of dropping it into the trash can. Disposal experts feel that this is just moving the pollution problem from one area to another, much as burning materials gets rid of solid wastes but adds pollution to the air.

Each year, 360 million tons of solid wastes must be disposed of in the United States alone. Recycling makes use of only a very small percent of these wastes. Some of the materials do not burn or decay. Some produce dangerous chemicals when they are burned. And almost all add to the problem of air pollution if they are not burned in modern incinerators. What can be done with all the trash and garbage?

Land fill is an answer for a small percentage of solid wastes. Sanitary materials can be buried in the ground each day and covered with a thin layer of soil to keep out insects and rats. Small amounts are being used in breaking roadbeds.

The dumping of wastes into lakes, rivers, and the oceans, which has upset the global environment, has been discussed in earlier chapters. Scientists must find new and better ways of disposing of the mountain of trash produced by the passengers on this crowded planet.

Perhaps by the year 2000, fusion power may vaporize wastes in such a way that reclaimed elements could be reused more readily. Scientists have been experimenting for many years to produce practical fusion, which is the combining of nuclei of atoms to produce practical amounts of energy. The atomic energy being used for peaceful purposes today is a process in which the nuclei of atoms are split. Taming the kind of reaction called fusion or thermonuclear power might help to reduce pollution in power plants, to desalt water, and to provide numerous benefits for mankind. Unfortunately, the process requires temperatures of hundreds of millions of degrees, temperatures which melt ordinary containers. Still, progress is being made and many scientists hope that by the year 2000, a fusion torch may help to keep man from drowning in his own waste.

In the meantime, search for other disposal methods proceeds on many fronts. Squeezing, or compacting, solid wastes is becoming more and more popular as a method of disposing of them. Machinery can compress tremendous amounts into small odorless bales which are dumped as sanitary landfill in areas not far from some cities. For example, in Minneapolis, one company uses 2,800 pounds of pressure per square inch to crunch garbage into bales, which it then trucks seventeen miles away and dumps in an area that can be grassed over in years to come and used as an industrial park.

In many cases, new methods of disposing of solid wastes are examples of men running just to stay in the same place; for as fast as they are developed, the amount of solid wastes and garbage increases. What happens to the garbage and solid wastes in your community?

When you consider that solid wastes are increasing about four times as fast as the population, and packaging wastes are increasing six times faster, you can see that something must be done so that man is not knee-deep in trash and garbage. Something must be done with sewage besides dumping it into the lakes and the rivers and the oceans. Better ways of recycling, better sewage treatment plants must help America dig out from under the mountains of trash that are accumulating. There are better ways of disposal than dumping, but this is still the primary way of getting rid of refuse. The odors, the flies, the roaches, the rats, and the material that runs from the dump into the ground and eventually into the surface water at some dumps make it obvious that better ways are needed. The situation is so bad at present that many people refer to it as the solid waste explosion. New solutions to this problem must be found quickly.

You can help as a pollution fighter by finding out what is being done in your community about sewage, chemical pollution, disposal of solid trash and other waste materials. You can write to companies that package their products with more paper and plastic than necessary. You can use bottles that are returnable and can be used again. For example, a glass bottle may make thirty-one round trips before it is disposed of in a waste pile. Your help is needed to see that Planet Earth does not become a global dump.

The day may come when solid waste is considered a resource to be mined rather than something that must be dumped, burned, or disposed of in other ways. Some researchers are exploring the possibility of recovering valuable metals, including gold, which are found in the trash heap. Mines have already developed techniques for separating some metals, such as tungsten, and recovering nickel that is used in electroplating. Junk piles of automobiles have long been an eyesore and a problem. One approach to disposing of old cars is the use of a machine that turns the old automobile body into a lump of solid metal through the use of pressure and heat. Much experimentation remains before automobile junk piles disappear and become practical supplies for tomorrow's use.

Although you can never really throw anything away, for everything on Planet Earth remains here even though its condition may be drastically changed, recycling still has not revealed its potential to man in the management of wastes.

Many young people are showing concern about the quality of their environment and are crusading for the production of materials that will last longer and will not create disposal problems.

Certainly, in addition to improving methods of recycling, and methods of compacting trash so that it will take less space in disposal, men must take a long look at the question of living in an environment that has a limited amount of space for waste. Nature can renew itself with time but not if man covers the earth with his trash.

8 LAND FOR TOMORROW

Someday, there may be just one acre of soil per individual in the world. By the year 2000, the number of passengers aboard Spaceship Earth may well be twice the number of passengers today. Science and technology can help the physical problem of overpopulation to a point. Even if enough food can be produced or manufactured, every trace of privacy, of freedom, of dignity, of adventure may well disappear.

Suppose those who are working for population control succeed to some degree. Even then man must search for new living space, for he is already in a world that is far too crowded. He already lives in a world where the soil and food are too scarce. Over four million people may die of starvation this year and large famines are predicted for future years.

When you fly around the world, if you do not look

closely, you may wonder why people are complaining about the lack of living space, for vast areas are unpopulated. At one time, when man dirtied his environment, he packed up and moved to a new area. The vast unpopulated areas of today are not desirable ones in which to live. Two-thirds of the passengers of Spaceship Earth travel in the tropic and subtropic regions of the planet. Only about 29 percent of the surface of Spaceship Earth is dry land, and of that only about one-tenth is suitable for agriculture.

Pollution fighters would gain much if men could spread out on more of the earth's surface. People talk of living beneath the sea, of living on the moon, or of living in caves in the earth. Such living must be far in the future.

Some people are building homes on land fill and extending the edges of the land into the water; others are pushing farther into the desert as a result of irrigation; and others are climbing higher up the mountains. But the passenger space of the planet is not being increased at a rate that in any way compares with the increase in the number of people.

With modern climate control, you might consider moving to the tundra, the land area along the margins of the Arctic Ocean. Here you would find unspoiled beauty, privacy, and wide-open spaces. But where would you get your food? Although Eskimos, Lapps, and Indians exist in this area, there is little hope of any large-scale farming on the tundra. One would not expect to find much life in an area that has just about eight inches of rain each year. Some deserts have this much. But temperature in tundra country makes this ecosystem far different from that of a desert. During much of the year the temperature is below freezing, with an average winter temperature of 10° below zero. Most of the year the water is frozen, but during the summer months it melts. In July and August the average temperature rises to about 40°. This is true at the surface; but a layer of ground known as permafrost remains frozen all year around. The melted snow, which cannot sink below this layer, forms pools of water and supplies the life of the tundra. The fragile green mat of plants, such as algae, fungi, and mosses, covers the ground. But this is not farming country.

This fragile Arctic ecosystem is one of great concern for pollution fighters and ecologists. It is one of the few major systems left unspoiled on Planet Earth. As men invade these far reaches in search of more oil for more energy and other natural resources, they must cope with many problems. Planning must be long-term, for haste can be dangerous. Imagine the difficulty and the dangers of laying a pipe four feet wide, full of hot oil, in a bed of permafrost. Such a giant pipe may rise, slide sideways, or fall deeper into the ground, depending on the conditions of the soil. It must twist and turn over 800 miles. If it snaps due to strain, millions of gallons of oil could be spread through the spongy Arctic ground. No wonder the question of pollution arises here.

In one attempt to prepare for the oil pipelines that will bring oil out of Alaska, companies are experimenting with different kinds of winter-hardy grass and special fertilizers to help the grass grow. When man breaks the green mat of the tundra, such grasses could help the ground cover renew itself. Otherwise heat from the long periods of sunlight would turn the frozen mud into a swamp. The water would run, and the erosion which it caused could never be repaired. In this and in many other ways man must look ahead to avoid mistakes of the kind which were made in the past.

As man looks out to other parts of the earth for living space, he finds 20 percent of the dry land is flat, grassy, and marshy. These grasslands are known as savannahs. Their soils are generally poor, they have few trees, their use for growing food is limited.

Suppose you look to the tropics for more living space and a place where food can be grown. Twenty percent of the dry land of the earth is luxuriant tropical rain forests, such as the Amazon Basin in Brazil, much of Central America, Indonesia, and the Congo Basin of Africa. Sometimes these areas are so overgrown with plants that a person cannot walk through them. Poor soils and heavy rainfalls, combined with high temperatures, make them uninviting places in which to live and almost impossible places in which to grow food. But people of the crowded Planet Earth are considering such areas out of necessity. As much as a billion acres on the edges of the rain forests might be changed so that food can grow there. Although these vast areas are low in fertility for crop plants, man may find ways of making them fertile.

The search for space to live and for more soil on which to grow food is full of complications. Suppose one cuts large numbers of trees in the tropical forest to make a home. The floor of the forest has been protected from the sun for thousands of years. The decay which goes on unnoticed provides food for the trees and plays a part in the balance of nature. When the trees are cut, the soil is exposed to the sun and it dries out rapidly and blows away.

Plants help to form soil and are part of the rock-soil cycle. If you live in the city, you might notice little plants between the cracks in the sidewalk that are pushing it apart or cracking it. If you live in the country, look for large rocks with lichens growing on them. These are flat greenish or white patches. Scrape off these tiny plants and look for bits of rock that may chip off at the same time. Look at the rocks under the area where the lichens are growing. Since they form a weak acid, you may see whitish streaks there. This acid weakens rock and plays a part in breaking it into smaller pieces in the process of soil formation. As you can see, making soil is a complicated process. It is also an extremely slow process.

Although part of all soil is rock, soil does not all come from rock. If you were to write a recipe for soil, what would you include? Here is an activity which might help you.

Take an aluminum foil pan or pie tin and spread spoonfuls of soil in it. Break it into small bits with your fingers. Now look at the tiny pieces through a magnifying glass. In addition to bits of rock or pebbles or sand, what else do you find? Can you see pieces of leaves and roots or other parts of plants? Look for small insects and bits of earthworms.

Some soil animals are very, very tiny and cannot be seen with the naked eye. Earthworms are the largest animals that dwell within a few inches below the surface of the soil, and they have been known to grow as long as six feet. If you could count the number of earthworms in one pasture in a temperate part of the world, you might find a great many—so many that, if they were all put on a scale, they would weigh as much as a large cow. These animals help the soil, as do many of the others which break it apart and make air spaces that are helpful for plants. There are roundworms in the soil, and various other kinds with which you are probably not familiar. In addition to this, there are numerous insects and their families. It has been estimated that one square yard of soil may contain as many as a million insects and their close relatives. And added to these, there may be more than a million worms and microscopic animals. Most of these soil animals are so small that if all were collected from an acre, they would weigh about a pound.

Here is an activity that will help you recognize the differences in quality of the soil. Take a walk, and carry plastic bags and a shovel, to collect samples of different kinds of soil. You may be able to find clay soil, which is very fine and light in color; gravel soil, which is coarse and has pebbles in it; loam soil, which is a mixture of sand and clay; humus soil in the woods, which is dark and loose and includes a large percentage of decayed plant and animal material. Sandy soil can be identified by the small grains of sand which you can see when you look through your magnifying glass. Clay soil can be identified by the sticky quality which it has when water is poured on it.

All of these soils are commonly found in gardens or farm areas where people live. If you traveled farther, you could collect many more kinds of soil. But much of the soil that is found near the edge of the desert or in the cold parts of the world or in the jungles is not of much value for the growing of food. This is why pollution fighters are recognizing the value of soil that is good for growing things and are trying to encourage people to save this kind of soil.

Here is a project for a group of conservationists. Take a walk through your neighborhood, and look for places where soil is being lost unnecessarily. Look for bare hills with gullies running up and down. These will become wider and deeper with each rainstorm, and the soil will run away to be lost for years to come.

Can you find hills where the furrows go up and down? Compare the amount of soil that runs off these with soil loss from hills that are terraced in circles running around the hills, rather than up and down.

Look around you for places where running water has caused erosion of soil. Try to find ways of preventing further erosion.

Here is an experiment that will help you to understand how water plays a part in the erosion of soil. Use a sheet of paper about the size of a piece of typing paper and paste it on a piece of cardboard the same size. A piece of white cardboard will serve the purpose, if you have such a piece. Lay the cardboard flat in the sink or on the floor or on an old table. Drop colored water on the paper from a medicine dropper, and notice the size and shapes of the drops that fall on the paper. Now prop one end of the cardboard so that it is about six inches higher than the other. Drop colored water from the same height as before and notice the size and shape of the splashes. Vary the height from which the colored water is dropped, and vary the slant of the cardboard. Think how this applies to raindrops falling on flat land and sloping land, and how it applies to gentle rains and hard rains.

In this experiment you can note the effect of rain on loose soil. Half fill three different coffee cans or flower pots of equal size with loose soil. Press the soil gently into the containers, and place some coins in similar positions on top of the soil in each one. Place each container in a sink or basin and sprinkle water onto each from a sprinkling can or a tin can with holes that have been punched into the bottom so that water will sprinkle through them. Sprinkle the water gently on the surface of the soil in the first container. Sprinkle more water harder to represent a heavier shower of rain in the second container. In the third one, simulate the effect of a heavy rainfall. Note the difference in the amount of unprotected soil that is splashed away from the coins. Here is another example of the erosion caused by water on soil.

The trees of the world have helped to improve the soil for millions of years. But the forests are vanishing. Once they were considered a nuisance to those who wanted to farm or build homes. They were in the way of highway builders and men who laid railroad tracks and communication wires. Forests were burned or cut down, so that many places turned from fertile green to dusty brown. Deserts have crept over much of the land that was once fertile.

The forest plays many important roles, only one of which is preventing the drying out of soil. Some scientists believe that as much as 1,000 million acres of the 4,000 million acres of forest that have been destroyed by man could be replanted, and on these acres forests could once more provide the important protection that civilizations need in order to flourish.

Perhaps you can join a group that is helping to save the soil in your community, or encourage a group to which you already belong to adopt a project that will help to save the soil. Many of the books listed in the Bibliography will give you suggestions for projects in this kind of pollution fighting. Just a few follow.

Look for soil that has been washed across a road. See if there are ways in which this can be prevented.

See if there are places in your community where pollution fighters can introduce any of the following improvements:

1. More planting to hold the soil
2. Reservoirs in the hills to catch and hold water for later use
3. Building protective walls along rivers where they overflow their banks
4. Building check dams of earth, rocks, or sticks in gullies or mountain streams

About a million acres of land in the United States are bulldozed each year for housing, airports, highways, business, and other construction. Trees and all other forms of plant life are uprooted. Whole communities of squirrels, rabbits, reptiles, and other animals are displaced. The soil creatures are destroyed without notice as the bulldozers roar and rumble over the area. Treads grind into the ground; trees topple as enormous earth-moving machines press against their trunks.

Man has been called a diligent destroyer. Strip miners have devastated large areas. It has been estimated that "industrially disturbed land," land stripped of its natural ground cover by strip miners, equals the combined areas of Connecticut and Rhode Island, and since the demand for coal is rising so fast, many more acres may soon be stripped. Certainly, man needs energy. Does he need wilderness?

In the United States, many who agree with Thoreau that: "In wildness is the preservation of the world," are striving to save the small amount of wilderness that is left. Once it is gone, it can never be replaced. Such living heritage is a priceless possession. VOICES FOR THE WILDERNESS is a report of the growing movement to preserve the wilderness. You will find this paperback book listed in the Bibliography at the end of this book.

Today's pollution fighters are trying to make ecologists part of the team preserving the wilderness, digging new waterways, and building new dams in projects around the world. Without far-reaching plans that include a thorough knowledge of the earth, the plants and animals, and the way they are related to each other, men sometimes build only to find that they have created new problems such as in the case of the Aswan Dam mentioned earlier.

Irrigation, which helps the farmer in many ways, can also create problems. Every drop of water brings with it a little salt, and as the water evaporates the salt remains. After a period of time, the salt gradually poisons the soil and the plants. There has been large-scale damage in Pakistan as a result of improper irrigation. Scientific knowledge has taught farmers how to solve this problem of too much salt by the use of drainage and spreading the water through the land in a process known as leaching. This is a percolating action. Large numbers of people must work for a long-term success.

Learn about projects in your community which may lead to large-scale changes in soil and land area. Check with local branches of some of the organizations listed at the end of this book.

9 FIGHT FOR CLEAN AIR

Picture the emergency waiting room of a local hospital. Some people are groping for space on the benches as they gasp for breath. Others are coughing and choking. Within the past few hours, 120 patients have entered this part of the hospital with respiratory problems due to the increase in the smog.

Such scenes are becoming increasingly common as the air becomes less clean. In New York City, it has been estimated that over 203,000 tons of dirt and poisonous chemicals fall into the air in one year. Each minute you are inhaling about 18 times. If you live in New York City, about 70,000 particles of dirt and a number of unhealthful gases go into your lungs with each breath. In Los Angeles, the smog has long been a serious problem, although some improvement has been accomplished.

In cities around the world people are showing growing concern because of the increase in disease due to smog. Even the polar bears in zoos are suffering from diseases such as lung cancer, probably a result of polluted air. As far away as Greenland, the lead concentration in the snow is 500 times more than at the natural level and this has blown there from polluted air of cities far away.

The statistics are bleak. Not only are people suffering in greater numbers from diseases such as emphysema, a condition in which lungs lose their capacity to function, and lung cancer, but there are increases in other diseases because the "garbage" in the air is increasing. Bronchial pneumonia, asthma, heart disease, and chronic bronchitis are among them. Certainly, pollution fighters have a tremendous task, for they must prevent these pollution-caused diseases.

In cities, there are people who are sealing their houses from the polluted air outdoors and installing filters to purify the air before it circulates through their homes. Just as many people used to move to better areas when the soil was depleted around their homes, today some people find the situation in the cities so serious that they are moving farther from the cities for fresher air. One hears talk of the fresh, pure air in the mountain areas and at the seashore; but even there air is suffering from pollution.

Just what is air? Air, by definition, is what is in the atmosphere. What does it contain? Chemists list the parts of the air in the first 60 miles above the surface of the earth as follows:

Recipe for Air	*Percentage by Volume*
Nitrogen	78.088
Oxygen	20.949
Argon	0.93
Carbon dioxide	0.03
Neon	0.0018
Helium	0.0005
Methane	0.0001

All of these vary somewhat from time to time and place to place.

Water vapor varies from zero to 4 percent. If all the water vapor became liquid and dropped to the surface of the earth, it is estimated that it would form a layer only one inch thick. Obviously, the water vapor in the air varies in concentration in different places because of the motion of air masses. While some places are having as much as 40 inches of rainfall in 24 hours, other areas are almost completely dry.

Air which contains only the chemicals mentioned

above is not found naturally. The air is always changing, and even in the country, far from industrial pollution, there are other ingredients in the air. Many of these are quite desirable. For instance, you may smell the gases which are given off by the sap that rises in trees in warm weather. Numerous plants add chemicals to the air, and many streams may be recognized for their private blend of oils, depending on the plants that grow on their banks. This is one explanation of why homing fish, such as salmon, are able to return to the exact stream in which they were spawned. The vapors from the air in that area get into the stream. The salmon's delicate sense of taste helps it to find its way, even after the fish has spent two years in the faraway oceans.

Many plant pollens and other chemicals which come from plants are disagreeable to man, and some cause hay fever and other allergies. But, in general, the air in the country contains relatively few pollutants that harm man unless "industrial parks" are nearby.

In the cities, on some days gritty black particles sift down from the dark skies. These are just one aspect of air pollution, known as particulate matter. Since our lungs are lined with thousands of lashing, hairlike cilia that beat back most of the invading particles and prevent them from entering, the human body can usually cope with a fair amount of particulate matter. Some does slip right through, and lodges in the lung linings. This makes it more difficult for lungs to function. The more particles that invade the human body, the more difficult it is for the lungs to carry on their normal exchange of gases.

In addition to the particles that fall from the skies, invisible poison gases and acid mists are often part of the pollution in large cities. Some of these gases are: sulfur dioxide, nitrogen oxide, ozone (a form of oxygen), and carbon monoxide. The levels of carbon monoxide in heavy traffic are sometimes so great that this poison can affect human judgment and behavior. This factor in air pollution, along with the darkening of the air by pollution, probably contributes to the number of automobile accidents.

Much research is being done to find ways of making automobiles practically pollution-free, since they contribute such a large amount to the air pollution in many areas. Some automobiles that are being produced will operate on lead-free gasoline, which is being developed in an effort to reduce harmful pollution by automobiles.

Lead pollution of the atmosphere is alarming many scientists who are studying changes in today's air. Dr. Tsaihwa J. Chow of Scripps Institution of Oceanography at La Jolla, California, has been studying the lead content of the air over the San Diego area for a period of three years. He reports that on some winter days the so-called safe limit, which government agencies have proposed, was only 20% greater than the lead content which he found. Dr. Chow is alarmed because the general level has been increasing at a rate of 5% a year. No wonder those who are fighting air pollution are pushing for lead-free gasoline for automobiles!

The United States has established standards for air quality, which make it necessary for pollution from automobile exhausts to be reduced through the coming years. Pollution fighters must see that such regulations are enforced, for there is some evidence that lead and other pollutants such as nickel and cadmium may contribute to the increases in cancer.

The airlines of the United States have voluntarily agreed to control the amount of particulate matter that

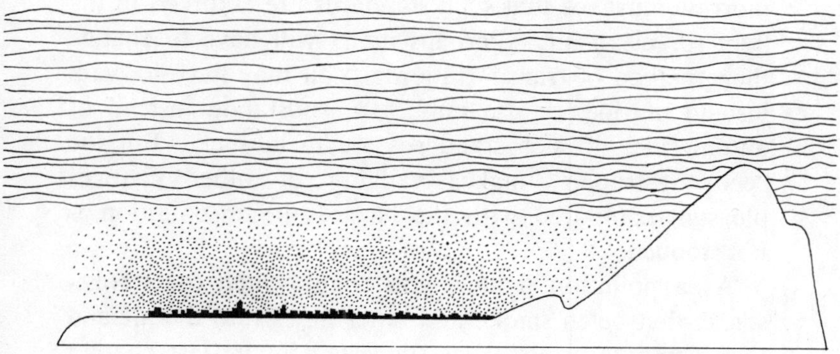

spews from jet planes. They are working in many directions to try to reduce pollution of the air, to which they contribute about 1 percent. This figure is misleading since pollution is concentrated near airports and lower in areas where fewer planes fly. But efforts are being made to control such pollution.

Air pollution is a very complicated problem. In one year, 200 million tons of poisonous materials get into the air of the United States from industries, city incinerators, apartments, homes, and such simple things as the burning of leaves in a garden. For each person in the United States, there is close to a ton.

In addition to its great hazards to human health, air pollution damages buildings, crops, fabrics, farm animals, and many other things. Although controlling pollution may seem expensive at first, clean air is a necessity.

In the United States, the Environmental Health Service is presently carrying on a nationwide effort to restore clean air, which includes participation of state and local people in large cities. The pollution fighters of tomorrow must see that such standards are enforced in the fight for clean air. Perhaps many will have to change their method of transportation. You may have to walk instead of ride, or use public transportation instead of automobiles, if such standards are to be met. But the "eco-catastrophe" makes walking or other changes pleasant compared with the hazards if air pollution is not stopped.

As a pollution fighter, you could watch for smokestacks that belch smoke and other gases into the air and take community action as suggested in the last chapter of this book. You can try to prevent smog which is so bad that it irritates eyes, and causes diseases related to breathing.

Even though you live far from city smokestacks, you cannot consider yourself safe from the growing menace of air pollution. Air does not stay put. Even when air seems to be completely still, it is in motion. Temperature changes and the turning of the earth cause the wind, which is part of the brewing of weather.

Since warm air takes up more space than an equal amount of cold air, warm air is lighter and rises. Heavier cold air moves in to fill the space left by the rising warm air. This movement of air is wind. So the wind blows, sometimes gently and sometimes with tremendous force. And air moves from place to place. It carries substances over long distances, around the world in due time, and from the North Pole to the South.

You can show that air expands when heated by a simple procedure. Just attach a balloon to the top of a soda bottle which is "empty," or actually full of air. Put the bottle in a pan of hot water and observe to see if there is any change in the size of the balloon.

Weather plays yet another important part in the

amount of pollution. If you have ever climbed a mountain, you know that the air gets cooler when you get higher. Sometimes, when the surface of the earth cools quickly, and the air next to it is also cooled, a layer of warm air just above this bottom layer near the earth becomes sandwiched between two cold layers. Actually the cold air pushes under the warm air and lifts it up. This is called an inversion. The warm layer acts like the lid on a pan, and prevents the natural rising of the air beneath it. Smoke from stacks and automobiles cannot be mixed with the air above, so the amount of pollution piles higher and higher and is sometimes extremely dangerous, causing many deaths.

Sunny weather adds to the problem of air pollution when certain gases are present. They react with each other in the sunlight to produce new compounds which make visibility more difficult and produce eye irritation, as well as crop damage and other problems. This is known as photochemical smog.

The effect of air inversion on human health has been known for some time. Back in 1930, a thick fog covered much of Belgium. Thousands of people became ill and about 60 died. Scientists believe that the fog was caused by sulfur oxides, gases that were produced in this heavy industrial area, and by the weather conditions in which the air was especially stagnant in a fifteen-mile stretch along the Meuse Valley.

You can illustrate inversion in the following way. Find 2 jars that are the same size. Place one in the refrigerator or in a pan of cold water and allow the other to remain full of air at room temperature.

After the air in the first jar has had a chance to cool, quickly insert some smoke into it by using punk, wet burning paper, or rope. Immediately, before the smoke escapes, invert the warm air jar over the cold air jar. Why does the smoke stay below?

Now show what happens if there are normal conditions. Use two jars as above, one with warm air and one with cold air, but add the smoke to the warm air and set up the jars with the cold air on top and the warm air on the bottom. Normally the warm layer of air would be near the earth with only colder air above. Can you see the smoke swirl up? Compare this with weather conditions which are normal and those which produce an inversion. No wonder weathermen play an important part in air alert systems.

You can start your own pollution watch by keeping a record of the distance that you can see each day. Use landmarks such as high buildings, hills, water towers, and the like. Keep your record over a long period of time and mark the farthest distance that you can see at a certain time in the morning and in the late afternoon. Do you think the air pollution problem is getting better or worse in the area where you live? Are there reports by government officials in your community with which you can compare your estimates?

Air pollution problems are not new, but in many places they are becoming increasingly worse, and have many dangerous implications.

There is much to be learned about the effect on climate from man-made dust in the atmosphere and from carbon dioxide buildup. The first has a cooling effect, the latter a warming effect. Man's knowledge of what is happening to the air around him and control of what he is doing to it may be fundamental to the future of the atmosphere, and in turn to the future of man. The information that is missing is obviously *important* information.

The carbon-oxygen cycle may have a serious effect

on the world's climate within the next thirty years, according to many scientists. They do not agree as to whether the climate will become cooler or warmer, because two effects are operating in opposite directions. The amount of carbon dioxide and particulate matter, or dust, in the air tends to trap the heat of the sun near the earth and to make the climate warmer. On the other hand, particulate matter reflects light back from the sun into space, and this tends to make the climate cooler. Both of these effects due to air pollution are increasing, and may continue to do so as population and industrial wastes increase through the next thirty years. Certainly man needs to learn more about how his activities affect climate before a serious change takes place in one way or the other. A rise of less than 10 degrees might melt enough glacial water at the poles to cause severe damage along the coasts. Cooling could cause damage to living things, or perhaps a new ice age may come if the cooling is great enough.

Cities may trigger their own snowstorms because of particles in the air as a result of pollution. One report which was made to the Environmental Services Administration indicated that a big city like Buffalo produces snow much the same way as heavy cloud seeding does. The dust particles which rise with the warmer air produced by such a large city act like the seed particles which scientists use in their cloud seeding techniques. Certainly men around the world are beginning to realize that they must do much more ecological homework before they understand the effects of air pollution.

What can you do about air pollution? First of all, consider yourself as a polluter as well as a pollution fighter. If you burn leaves, you contribute to air pollution. If you burn trash in an inefficient incinerator either at home or outdoors, you are a polluter of the air.

If you are in an automobile, suggest that the driver turn off the engine while you are waiting for someone to do an errand or for passengers to come. Idling engines produce more carbon monoxide than those which are moving. Perhaps you feel that each bit of pollution, such as the ones just mentioned, amounts to nothing. But if each person contributes a bit, much accumulates.

Keeping house furnaces in good operating condition and finding out about the quality of air in your community are important tasks for pollution fighters. With more people, more industries, more automobiles, more trash being burned, more houses being heated in city and country, the situation will grow dangerously worse without great effort on the part of pollution fighters. The battle can be won only by the efforts of more people. Your help is needed.

If you smoke cigarettes, you are polluting your own air and the air of those around you. Many scientists believe that cigarette smoking causes more cancer than other types of air pollution. Many people are deeply concerned about the chemicals in the air which may play a part in causing cancer, as well as the various lung diseases mentioned earlier. There does not have to be an acrid stench or eye-smarting irritation to cause slow damage to tissues. In come cases, microscopic particles of materials that irritate or may be cancer-causing are present at peak level in the air on days that are cool and refreshing and appear to be completely clear of smog or other air pollution. Many years may pass before the people become aware of pollutants in the air, unless they are concerned enough to find out about the results of analysis.

Through the National Air Pollution Control Administration, in the United States, and through various other agencies in countries around the world, extensive air-sampling networks are measuring the various air pollutants from factory smokestacks, home furnaces, garbage-burning incinerators, and most of all, from automobiles' exhausts.

In some cities, the task of cleaning the air is well under way. For example, back in 1941, in Pittsburgh by noon, there were days when the sky was dark from particles polluting the air. Although much of this matter has been eliminated through control efforts, there are still invisible materials such as carbon monoxide and sulfur dioxide that must be reduced. Instruments called sensors are sniffing the air continuously and readings are made of the levels of pollution for various substances. These readings, along with information about the weather, which plays an important part in pollution hazards, are fed into a computer by telephone. The readings are printed at five-minute intervals. When a pollutant exceeds what is considered a safe level, readings are automatically increased, so that they are made at fifteen-second intervals. Systems such as this act as early-warning devices to prevent pollutants from reaching dangerous levels. When a rise is noticed, sources can be spotted, and action can be taken before there is a catastrophe.

Everyday automobile traffic may be drastically cut in some large cities as a result of The Clean Air Act of 1971. Emergency alert systems are going into effect throughout the United States.

It is your job to help maintain the standards which are now being set, and thus aid in preventing disease caused by pollutants.

No matter where you live, you can help by joining pollution fighters for clean air. Air pollution affects each person.

10 NOISE POLLUTION

The amount of noise around you is doubling every ten years. The sounds that batter your eardrums also challenge your spirits. This is especially true if you are a city dweller, where you are exposed to a multitude of unwanted sounds. Repair in streets and construction of new buildings is accompanied by loud noises from air compressors, wrecking balls, jackhammers, pile drivers, and riveting equipment. In addition to this, car horns blare, trucks rumble, sirens wail, subways screech, and planes roar.

Pollution fighters and health experts are greatly concerned about the effect of too much noise from rock music, motorcycles, airplanes, garbage trucks, and an almost endless list of noise producers that are already causing permanent hearing damage to many.

Although quieter air conditioners, lawn mowers, and

other machines are usually somewhat more expensive, the cost to society from excess noise is tremendous. The World Health Organization estimates that industrial noise alone costs the United States more than 4 billion dollars a year in absenteeism, inefficiency, accidents, and compensation claims. In addition to this, the mental upset caused by noise cannot be measured in dollars.

What happens when a loud noise enters your ear? Medical researchers have shown that too much noise narrows the arteries that carry the blood through your body. It increases the heartbeats and dilates the pupils of the eyes. One study suggests that noise now causes 70 percent of nervous upsets in one large city, compared with only 50 percent four years ago. Such statistics are hard to prove; but few disagree that noise pollution may be one reason for an increase in heart disease and mental illness in many areas. Some medical authorities feel that noise pollution has helped to increase allergies, stomach ulcers, indigestion, spinal meningitis, the amount of cholesterol in the arteries, and numerous other problems.

As everyone knows, noise is caused by vibrations, and it differs from music and more pleasant sounds in that it is "unwanted sound." In addition to the general upset mentioned above, noise can cause deafness by destroying microscopic hair cells in the inner ear. Such cells play a part in stimulating nerve cells which transmit electric impulse to the brain. One very loud blast may destroy thousands of cells, and these cells never regenerate. If one is constantly exposed to lesser noises, there is a slower destruction of the hair cells.

Most people who live in today's cities have already lost a large number of such parts of their hearing mechanism, but they probably don't miss it because nature has supplied them with an excess. It is possible that the loss of hearing which has often been attributed to old age is due to excess noise. Dr. Samuel Rosen, a famous noise pollution expert, has made numerous studies on the relationship between noise exposure and learning ability. Where living is quiet, hearing is better, especially in the aged. Studies in the Sudan, Africa, where a tribe of people live quietly without even the noise of drums, seem to bear this out. This tribe, the Mabaan, have extraordinary hearing, and no heart disease or high blood pressure. Dr. Rosen studied these people on three different trips to the Sudan over a period of many years and has made noise pollution studies in other far parts of the world. He believes noise is a definite contributor to heart disease.

According to a recent report by the Surgeon General

of the United States, between 6 to 15 million Americans are going deaf from occupational noise. Some scientists believe that if city noises continue to rise at their present rate, many people will be deaf by the year 2000. Certainly pollution fighters are needed to turn down the noise.

Before one can talk intelligently about noise control, he must learn something of the yardstick that is used to measure noises. The *decibel* is a measure of noise based on the sound pressure that moves the ear. The weakest sound that can be picked up by a healthy human ear is zero decibels. A faint breeze might rate zero decibels.

How much noise can you tolerate? How much noise do you hear? What is the real meaning of a decibel? Measuring noise is difficult, for even though one uses a decibel as a unit many other factors are involved. For example, high-pitched tones are more annoying than low-pitched tones, so some scales give less weight to low tones. The letter *A* is sometimes seen after the letters *db*. *A* refers to pitch; *db* to amount of noise, or decibels. Ear injury may begin at sound levels above 80 dbA.

One reason that decibels are so difficult to understand is that they do not increase in ordinary one-two-three sequence but are logarithmic. This means that ten decibels are tenfold one decibel, twenty decibels are one hundred times the power of one.

Listen to some of the noises around you and compare them. You might start in the kitchen of your own home. Suppose noises from the dishwasher, garbage disposal, food blender, rattling pots and pans are occuring at one time. Activity such as this in a kitchen has been measured at 93 decibels.

Here are some other examples of noises measured in decibels:

1 motorcycle	90
1 garden-size power mower	98
1 window air conditioner	55
a city school playground	78
some rock or soul music	107
1 jetliner 500 feet overhead	115
an average street	65
a busy street	85

A factory worker in a print shop may be exposed to 95 decibels each day; and an average of 118 decibels each day if he works in a factory where boilers are made. Many construction workers live with more than 100 decibels during each working day.

Although the point of discomfort varies from 80 to 115 decibels with various people, and pain begins in the 115 to 145 decibel range, everyone agrees that there is such a thing as noise pollution in much of the world. According to the Department of Labor in the United States, 90 decibels is the longest continuous noise that one should be allowed to tolerate in an occupational setting for a period of 8 hours, day after day. This does not mean that such noise is safe. When you consider that the decibel scale increases in such a way that 150 decibels is one million billion times as great as one decibel, you can see how important the decrease of just a few decibels may be.

What can you do to help control noise pollution? One method is soundproofing your home. But this is expensive and it helps only you, and only when you are inside your home. Campaigns against horn-blowing have been successful in some countries. In New York City, a task force on noise control reported that a three-year study disclosed that the noise level had reached such a high degree that it was a peril to city life. The report called for rules to eliminate disturbing noises and included a city ordinance to prohibit excessive truck noise. Quiet garbage trucks have been developed in cooperation with General Motors, and plastic and paper bags are being used in some areas rather than metal trash cans. But the elimination of noise from cities such as New York, and even from smaller cities and country areas, is a complex problem. For example, the bags increase the amount of trash.

In 1970, new noise regulating provisions were added to those already limiting the amount of industrial noise in plants that do work for the U.S. government. European building codes have long had noise-control clauses. Subways in Montreal, Paris, and Mexico City use rubber wheels to help dampen sound. These are just a few ways in which people are trying to help control noise pollution. But progress is slow, and much more needs to be done on many fronts.

Various studies show that even though the human body is able to adapt to many changes in the environment, it does not become conditioned to noise. You can control some exposure to noise. For example, you can limit the amount of time you listen to rock music and avoid using headphones with sounds at high intensity that will damage large numbers of irreplacable hearing cells. But what can you do about the noises on the highways and in the skies?

The Federal Aviation Administration has authority to fix aircraft noise limits, and efforts are being made to lessen the amount of noise pollution. New jets must meet certain standards. Aircraft manufacturers are working to reduce the high-frequency whines of the jet's fans and the low-frequency roar of the jet's exhaust.

Since the roar is loudest when planes take off and maximum thrust is needed, and the whine is loudest when the plane is making its approach to land at an airport, it is the people who live and work near an airport who are disturbed most by noise pollution from planes. A reduction of just ten decibels does not seem like much, but actually it means a 50 percent cut in an individual's awareness of noise. It also means a great expense for the people who build planes; for it is very difficult to make the planes quieter. But the new superjets, such as the L-1011 being built by the Lockheed Aircraft Corporation and the DC-10 of the MacDonald-Douglas Corporation, must comply with regulations making them less noisy than earlier planes. The new Boeing 747 is quieter than the Boeing 707.

What about the planes of the future, which will travel faster than the speed of sound? There has been much apprehension about the coming of the SST (Supersonic transport), which will produce a sonic boom. A sonic boom is a thunderous clap that is caused by shock waves of air. All airplanes compress the air ahead of them as they push through the sky. The faster the plane travels, the greater the pressure. But when planes travel more slowly than the speed of sound, the pressure disturbance moves out from the aircraft in all directions and is dissipated without affecting people on the ground. When planes exceed the speed of sound, the pressure area is forced out behind the airplane, where it becomes cone-shaped. Actually, one cone extends back from the nose and another cone from the tail of the plane as it flies through the air that it is disturbing.

Suppose you are standing on the ground as a supersonic plane flies above. When the bottom edges of the cones that it forms reach the earth, their waves are re-

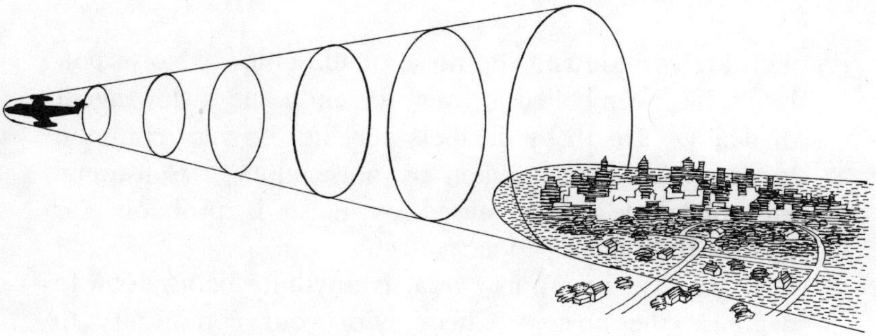

ceived by your ear as an explosive sound. Sometimes this is single, sometimes double. You hear this just briefly. But as the plane speeds through the sky, it continues to send out waves of the same intensity, so that people beneath it at various places along its path will be exposed to sonic boom.

Congress recently defeated a bill for further funding of the SST, and environmentalists played a large part in bringing this about. The foes of the SST claimed that 500 million persons in America, Europe, and Asia could be jolted every hour, day and night, by sonic booms if overland flight was permitted. Those who were in favor of the SST contended that they were optimistic that efforts would succeed in reducing the noise or in keeping supersonic flight over ocean areas, by the time these planes could be introduced in 1978.

Certainly, noise pollution, whether from supersonic jets, subsonic jets, or vacuum cleaners, garbage disposals, and other everyday appliances is something to be toned down. Many people are unaware of the seriousness of noise pollution and have felt that it was a necessary evil. Much can be done to reduce noise level, and citizens who are aware of its dangers can help to enforce antinoise laws.

The future of cities may depend on how successful

men are in reducing the noise of the cities. Noise pollution has been called a threat to sanity and a slow agent of death. Death by decibels may not be very common but men have been killed by noise alone. Emotional and physical illness caused by noise is probably far greater than most people realize.

If you live in a noisy area, is anything being done to decrease the noise? Check with your community to find out what you can do in this area, and contact some of the organizations listed at the end of this book for further information on action that you can take.

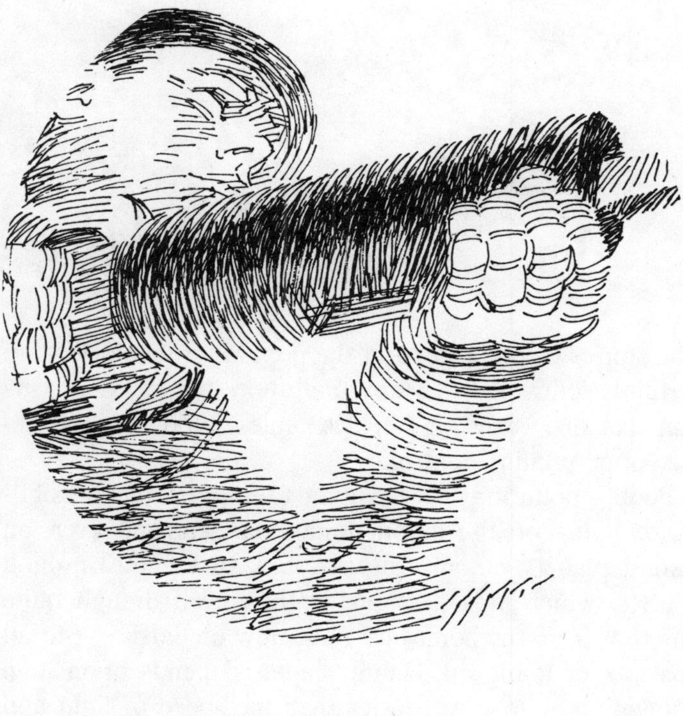

11 PEOPLE POLLUTION

The stories of pollution and the population explosion are certainly entwined. Many pollution fighters are certain that their cause is hopeless unless there is some degree of population control.

People pollution has led men to explore the possibility of living on the moon, under the sea, and even on distant planets. Some futurists talk of the 600 million planets, which may be quite livable, even though quite different from the homes we have now on earth. But all this talk of living on distant planets depends upon man knowing how to travel faster than the speed of light and science claims this is impossible. One cannot plan to buy a piece of real estate in the sky for living in this generation or the next. All of the planets that are as attractive as the earth lie far in outer space, perhaps far beyond Alpha Centauri, the nearest star other than the sun.

This star is four light years away, and since a light year is the distance that light travels in one year, or 5,878,-000,000,000 miles, you can see that moving to other planets is not a practical answer to the problem of a crowded and polluted earth.

Another "blue-sky" idea comes from a California physicist, Dr. Charles H. Townes. He suggests that if man were only half his present size and the earth remained the same size, living space, food requirements, and demands on natural resources would be considerably less. He does not seriously suggest a way of accomplishing this.

High-rise buildings that overhang water offer a somewhat more realistic approach to housing the already too-large population of Spaceship Earth. Living under glass or plastic domes is another serious approach. In areas where weather is extreme, such as deserts or Polar Regions, such huge domes might cover entire cities. Complete climate control therefore might help to solve the larger problem, providing it could be accomplished without polluting the outside air and disturbing the rest of the world. What would such living do to the psychological or emotional side of human beings? Would more trees and open spaces be available there than in the cities, where lack of open spaces is already producing stress?

Floating cities have been urged by some planners who have been trying to solve the problem of housing and congestion in coastal areas. Noiseless, trafficless cities, where much activity takes place underground, such as shipment of food and construction material, are being designed. Wastes would be removed by pipes connected to homes through vast plumbing systems. Whether man builds homes layer upon layer on top of the earth, with much of his necessary equipment and service material below him, or whether he modifies the environments and lives under the sea, or in Polar Regions, or in deserts, the rate at which man is reproducing himself still must be seriously considered. No matter how fast new plans are developed, the number of new occupants of Spaceship Earth is rapidly outspacing them.

While men dream of ways to help house and feed the increasing number of people who are filling the earth, more than two human beings are being added to the world's population each second. Look at your watch for one minute. During this time, at least 132 more people were added to the total population of the world. During a full day, 190,000 hungry babies are added to the staggering number of mouths that must be fed on this planet. In one week, the population of the world increases by 1.3 million. Since two-thirds of the people of today's world do not have adequate food and health care, and since pollution problems are already so severe, it is no wonder that more and more people are joining the crusade for zero population growth, or no more than two children per family.

At the present rate of growth, the population of the world will double in about 37 years. If growth continues at this rate, it will not be long before each person on the earth has about three or four yards of space to occupy. Long before such a time, many of the problems of supporting people on Spaceship Earth will become so great that your children and grandchildren will not be able to live in a world where they must be stacked body next to body.

Although at the present time there are eleven acres of land for every man, woman, and child in the United States, 80 percent of the nation's population is crowded into about 10 percent of the land area. An intensive study was made over a five-year period to determine what could be done with the one-third of the nation's land that is owned by the federal government. As a result, some public lands may be made available for the expansion of existing cities and for the development of new cities and towns. Certainly, the proposition of spreading the population across the nation deserves great attention and much action after careful study.

Maurice H. Stans, Secretary of Commerce of the United States, predicts that Americans will be jammed together in an "ant hill society" by the year 2000, unless government and business join in a coherent national growth policy. Problems such as crime, youthful alienation, congestion, and pollution will certainly increase sharply unless such crowding is avoided.

Even in the United States, where crowding is far less than in many parts of the world, many scientists believe that the optimum population level has already been passed.

Many people disagree with this. They feel that limiting population in countries such as the United States is not the answer to an overcrowded world. According to international standards, the United States is underpopulated. Many countries in Europe hold more persons per mile and have been able to maintain a lesser degree of pollution than the United States.

In many areas of the world, the increase in population rapidly outpaces the ability to produce food and other necessities of life.

How many passengers can Spaceship Earth carry? This is another question on which many people do not agree. The Green Revolution, in which new high-yield crops and meat substitutes have helped to feed a hungry world, has given some promise. Farming the sea is a hope of many who look to the future optimistically. But many experts believe that sea farms will not feed more than a small percentage of the hungry millions of the exploding population of Earth. The food supply, which does not keep pace today with the number of people in the world, may soon lag even farther behind. Some passengers who are traveling first class on Spaceship Earth may soon be feeling the pressures and pains of those who today travel in less comfortable parts of the planet.

Not everyone agrees with the doomsday prophets who say man may have less than thirty-five years to survive on the planet. But few people, if any, who see the quality of life in much of Calcutta, India, can do so without feeling some of the pain and tragedy of this overcrowded city. Picture Calcutta this very day. Whole families may never move from their homes on the sidewalks, the steps, and the gutters. Permanent street-dwellers range from 50,000 to a million, so that at night the rows of sleeping bodies look somewhat like corpses lined up one after another.

One of the better places to live in this city is called a buster. Such a home is made by erecting a roof four feet above the ground. Hundreds of thousands of people live in such slums, with an average of forty square feet per person. They wash in and drink highly polluted water. After a strong rainfall, the open-sewage drains spill over with a flood of human excrement. Medical care is practically nonexistent. Children play with the rats and feed them the way children feed squirrels in the parks of some other countries.

Many who study the problems of Calcutta believe that it is a dying city; for here, as in many rural areas, the predominant activity is a ceaseless struggle for survival. Daily, large numbers of people fling themselves from the bridge over the Hooghly River, rather than live with the despair and deprivation that each day brings. This same bridge is one over which 300,000 people struggle each day, looking for work or something better than what they have on the other side of the river.

Fifty-six miles down the river from Calcutta a new city is being built. Here the water is deeper, and shipping from the Bay of Bengal may provide jobs in petroleum refineries. Farther up the river at Calcutta, the water is so shallow that it cannot handle large ships. But the new city downstream may not be finished in time to save final breakdown of the sixth largest city in the world.

Calcutta is just one example of the problems of crowding on Spaceship Earth. Those who look to population control are concerned with human dignity and living conditions that allow each individual to develop his full potential. Birth control programs are being undertaken in many countries around the world. But in some countries, whole attitudes must be changed. For years it has been popular to glorify the large family. Much work remains to be done, and there are many organizations that are working hard against the problem of rising birth rates. If you wish to learn more of what you can do to help prevent or slow the population explosion, write to the organizations listed at the back of this book.

Today the great migration toward the cities continues, and urban planners work hard to try to solve the many problems of people living so closely together. Some feel that if a city such as Calcutta completely collapses it would be worse than a large flood or famine, for this might mean that the challenge of the city of tomorrow could not be met.

Certainly a world of "standing room only" would be a world of unspeakable pollution. Pollution fighting is a lost cause unless one limits the number of people who must travel on a planet with limited supplies and limited space for living and disposing of wastes.

Man is just beginning to discover that he must live in harmony with his natural environment to insure his continued survival and progress. Although he can alter the world around him through science and technology, there is still evidence that he cannot happily live in a world of concrete and steel. The artificial world that he constructs does not provide the same peace and joy that is found in the wilderness of nature. Man's well-being and spirit depend upon green trees and natural surroundings for renewal. All this is related to living space.

Urban conditions are part of the pollution problem.

Not only do they contribute to it, but they also intensify it. Clean air, sunshine, and unpolluted water must all be part of urban life if one is to avoid emotional stress in the city. Dr. Jerome Wiesner of the Massachusetts Institute of Technology has said: "We are engaged in a race between catastrophe and the intelligent use of technology, and it is not at all clear that we are going to win." Pollution fighters individually and in groups can reduce the risk of losing.

12 TOWARD A GLOBAL WARNING

Do you remember the stories of the miners who took canaries with them into the mines to serve as a warning system? The canaries showed the effects of odorless methane gas which might be present before the miners became ill from it.

Today there are many warning systems for many kinds of pollution. Voices are heard from around the world pleading for action by pollution fighters. Reports come from far and wide. For example, in France the problem of pollution in the Mediterranean is being studied. Here nine of thirteen species of food fish have disappeared since 1945. There are warnings that if industry and construction continue to increase, the continental shelf of France will lie beneath sterile waters. Sewage, which is dumped directly onto the sea, oil pollution, detergents, pesticides, asphalt, and plastic objects dumped

SYSTEM

by tankers at sea are all adding to the tragedy of a highly polluted Mediterranean Sea. Similar warning signals are being noted in the Baltic Sea.

From Athens, Greece, comes the warning that the city may have to be abandoned by its residents in ten years unless serious measures are taken at once to combat air pollution. The city will be covered by a cloud of industrial waste fumes that will damage the ancient marble of its temples and statues in addition to making the environment more unhealthful.

From Russia come the questions: Why do we see almost no flocks of geese and cranes in the sky? Why have the woods and gardens and fields become quieter? Where have the partridges gone? Pollution is a major problem in this part of the world, too.

The city of Venice has long been in danger from industrial combines that foul the air and the water. Recurring floods from exceptional high tides are increasing, and Venice as a whole is sinking into the lagoon inch by inch, year after year. While some people blame the serious situation in Venice on industry, others believe that land fills in the lagoon have upset the tidal system and increased the number of floods. Some say that new industrial plants on the mainland are pumping fresh water from the subsoil and upsetting the natural structure of the land and flow of the water. No matter what the reason, Venice is in danger. Here, as in America, young people have organized to help many other groups to restore the balance of nature and to prevent further pollution. Some say that it is too late; others try hard to save what is left.

Warnings come from many parts of the world on different varieties of pollution. In some cases animals are being used as part of a warning system, much as miners once used canaries to test the quality of the air in the mines. Goldfish in lakes are being observed, and on days when they seem less active, further tests for pesticides and pollution are made. Pigeons are serving in some areas as test animals in the field of air pollution. Hikers and bird watchers have long served as an early warning system of environmental change by recording the crises in wildlife. Some flowers are particularly sensitive to sulphur dioxide from smokestacks. Petunias are an example of this. Radishes have been used to warn of poisonous chlorine gas. Monkeys have been kept in cages to warn of yellow fever, since they catch the disease more easily than human beings.

There are many more scientific approaches to the international problem of pollution. For example, the Survival Service Commission of the International Union for Conservation of Nature and Natural Resources keeps the *Red Data Book*. This records the status of living things and changes that guide new policies in conservation and preservation of endangered species of animals and plants. The Commission reports that 100 of the world's higher animals have become extinct since the year 1600, and nearly one-fortieth of the remainder are now in danger. Naturalists, zoologists, geologists, ecologists, and conservationists, all over the world join together to provide information for the *Red Data Book*. WILDLIFE IN DANGER is a recent book for those who are interested in an authoritative account of the animals, birds, and plants of the world that are threatened with extinction. It is listed in the Bibliography at the end of this book.

Certainly, pollution is an international problem. Not long ago, "black snow" fell on Eastern Norway and Western Sweden. This was actually a grayish snow with black spots in it. It was discovered that the blackness was caused by pollutants that were rising into the air over the Ruhr Valley in West Germany. Normally these wastes are distributed throughout the atmosphere; but at this particular time they were trapped by what scientists call an atmospheric inversion. As you know, this is a situation in which the warmer upper air blocks the normal process of dissipation. The air blew the wastes first to the east and then back over Sweden and Norway.

Although cases of inversion are rare, air patterns of prevailing winds do contribute to pollution damage that knows no boundaries or borders. Sulphuric acid is a waste which is produced in the Ruhr Valley and in other industrial areas, as well as in Sweden and Norway. But some scientists feel that part of the damage produced by sulphuric acid in Scandinavian countries is the result of pollution from far distant places.

Sulphuric acid dissolves minerals and alters conditions in the soil. The minerals that are dissolved out of the soil drain into the lakes and rivers and upset the delicate balance of plant and animal life in these areas. Since the growth of forests depends on the richness of soil, experts are concerned about increased soil acidity. Countries must be concerned with pollution not only in their immediate neighborhoods, but also with pollution carried by prevailing winds from distant places.

Many international groups have already begun to study pollution problems. For example, an international team of scientists based in Paris plans to frame an overall prediction of the seriousness of air pollution problems. During a two-year period, twenty-two nations will take part in trying to learn more about air pollution around the world in order to make available the best antipollution devices.

Another program is the five-year project known as the International Biological Program, in which scientists of fifty-seven nations are learning more about the life cycle of all organisms in an environment of six different ecosystems. They are concentrating on ecosystems in several kinds of forests, in deserts, tundra, and grasslands. Eventually all the research findings from this program will be fed into a computer. Scientists from around the world will learn a tremendous amount of information that would be almost impossible to secure on a small scale. The computer could evaluate the consequences of eliminating a specific bird from a certain area. It could give information on what would happen if one particular kind of insect were to be wiped out. The results of this program should help men to help those species that can do most for a crowded and fragile planet, Earth.

While nations begin to plan for participation in the United Nations Environmental Year in 1972, other groups are calling for action now. In July of 1970, representatives from about fifty countries joined in urging management of the world's oceans based on participation by all nations. Their concern with contamination of the seas included pollution from chemical, industrial, and radioactive wastes. One participant pleaded that man recognize that "the time has come to face the terrible fact that our oceans and seas are threatened by destruction." He appealed to the world to save the seas and oceans for mankind. Many believe that if the oceans are not saved, mankind will not survive, since the tiny plants that grow by the billions on the surface of the seas play a large part in the production of oxygen of Planet Earth.

Zoologist Kenneth E. F. Watt of the University of California feels that more accurate knowledge is needed for the necessary planning that will save the earth from unbelievable problems. Along with other scientists, he has designed mathematical models of different kinds of ecosystems and is analyzing them by computer. The computers indicate that the world's oil reserves will be depleted around the year 2000. Coal reserves will be used next, and atomic energy may or may not be able to provide enough energy for the growing population of the earth. If not, zoologist Watt predicts that there will be severe starvation or wars to reduce the number of people from more than 10 billion to less than 4 billion. No wonder he urges population control.

While not everyone agrees with predictions or about how to control pollution, men around the world are becoming more and more aware of the need for an international network to warn of hazards in the environment. Man has learned that as he changes the world in which he lives, he changes himself. And he is more aware than ever that he lacks answers to many of the basic questions about the environment and the changes which are taking place.

Perhaps an environmental warning system will be the first truly cooperative effort of all mankind.

13 HINTS FOR POLLUTION FIGHTERS

People often ask, "Why doesn't someone do something about pollution?" When asked, "Who is the polluter?" they reply: "Chemical companies; automobile producers; the government; manufacturers who dump wastes into the streams and poisons into the air; users of atomic energy who spread radiation into the air, store radioactive wastes in the ground and dump them into the ocean, and do not consider the long-term problems of leakage; farmers who spray pesticides and add fertilizers to the ground which eventually run into the lakes and oceans and cause their early death." The list could go on and on.

You might try an experiment among your friends or classmates, asking them to make a list of polluters. Note how many people include themselves. Much pollution begins at home, for you cannot avoid being a

polluter. Along with the rest of your family, you contribute your share.

The mothers who wash the family clothes and dishes with detergents add each year 5 billion pounds of pollutants to the rivers and lakes of America alone. Members of your family who drive cars help to fill the air with toxic lead from additives, along with hydrocarbons, carbon monoxide, and nitrogen oxides. In America alone, they add about 90,000,000 tons of pollutants to the air. The grease from the ovens, the trash paper from the wastebaskets, garbage from your kitchen all add to the six pounds of solid waste which each member of your family produces each year. Most of this is either burned, in which case it adds to the pollution of the air, carted away and dumped at another place, where it adds to the pollution of the land, or dumped into our rivers or oceans where it pollutes the waters.

The gardeners in your family may still be spraying with "hard" pesticides that cause certain kinds of birds to lay soft-shelled eggs, so that the embryos never come out alive and the birds are unable to reproduce.

Suppose you come home from the supermarket with a load of food. You strip away the brown paper bags and the clear plastic wrappings from the lettuce and other vegetables; you toss away the wrappings from the toothpaste, the paper towels, and the frozen foods. You may have already noted that it does not take long to fill a large kitchen wastebasket. As you unpack, you put the soft drinks and other beverages into the refrigerator and eventually, unless you are using reusable bottles, these too add to the trash pile that comes from your house.

The grease, the bottles, the cans, the cartons, the paper, the pesticides, the detergents are all part of everyday life.

Today, as individuals, many pollution fighters are looking for ways to decrease the amount of pollution that they themselves produce.

BEWARE OF PEOPLE WHO CLAIM TO BE FIGHTING POLLUTION BUT WHO TALK MORE THAN THEY ACT.

Some pollution fighters do not see the total picture. They are first to criticize the existing pollution, without taking into account some of the reasons for it. Perhaps they have recently been in the hospital and enjoyed the comfort and cleanliness of the disposables which are used in many hospitals today. This is not just a luxury.

Because of the scarcity of help, and because hospital supplies must be sterilized, disposable sheets, hypodermic needles, and much other equipment have replaced the reusable things. Although this does add to the amount of solid trash, here is an example where there seems to be good reason for disposable materials.

Some pollution fighters see only the black side of the environmental picture. There is no question but that action must be taken, and taken fast. But as one looks at the environment, one must hope that in the years to come people will take better care of it so that the beauty that is present today can continue to be enjoyed. The Everglades still glow red in the summer dawn, the Hudson River still flows in a valley of beautiful green mountains, the coastlines in some places are still pure and free of dead fish and oil spills. There are still wide open spaces which people can preserve.

BEWARE OF POLLUTION FIGHTERS WHO TALK ONLY OF DOOMSDAY, RATHER THAN WORKING TOWARD A BRIGHTER FUTURE.

Today there is a great trend toward organic gardening, and this is good, providing one has knowledge. In a recent article in a widely publicized catalogue, a writer recommended the use of human wastes in raising one's own vegetables. A certain amount of experimentation and research is being done in this area, but one must be certain that the wastes are not disease carriers. The plagues of cholera, typhoid, and many other diseases were largely the result of the use of human-waste fertilizers. People in countries where sanitation has been raised to high levels have not built up an immunity to diseases that can be transmitted this way.

Putting the skins of canteloupes, watermelons, and such garbage in the soil is far different from the use of human and animal wastes that may be diseased.

BEWARE OF ADVICE THAT IS NOT BASED ON KNOWLEDGE.

Man shares the earth with more than a million different kinds of animals, and most of these belong to the group known as insects. Only about a thousand types of insects attack human beings, animals, and plants, but the damage they do in producing or carrying disease and in destroying food crops is almost beyond estimate. In the 1940s, with the development of DDT (dichlorodiphenyltrichloroethane), man seemed to have won the battle against insects. Malaria was brought under control by spraying and killing the anopheles mosquito in many parts of the world. The rat flea, which carries typhus, a disease of rats that also attacks man, was greatly reduced. Other diseases, such as cholera, bubonic plague, yellow fever, and smallpox were greatly reduced because DDT was such a potent insecticide. Crop losses due to insects dropped tremendous amounts. The insecticide DDT seemed like a miracle. BUT there were long-range effects. Not only was this an insecticide; in the long run it was what is known as a *biocide*, or life killer. Although the insects died within hours after DDT was sprayed on a crop, there were long-lasting effects on other living things. In the forests, in the soil, in the water that ran to the oceans, DDT was spread throughout the world. It has been found stored in the bodies of the penguins in Anarctica, in the tiny, one-celled plants and animals of the ocean, and to such a large degree in the fatty tissues of human beings that many feel that mother's milk is unsafe to drink because of the DDT content.

DDT is passed from one animal to another in what is known as the food chain. When one animal eats a plant or another animal, energy is transferred through the chain. The biocides such as DDT moved through the chain and became more concentrated as they passed from one link to another. Although the DDT did not always kill the animal, it affected reproduction in some cases and may have had unknown effects in others.

Man is the animal at the end of the food chain, and it has been estimated that the average person in the United States has twelve parts per million of DDT stored in his fat. So far, there seems to be no evidence that man has been poisoned by the accumulation of pesticides in the environment, although there are many questions and all the answers are not known. Some people believe that the accumulation of pesticides in the environment causes harm in the form of residues ingested in food and water, or inhaled when one breathes. The damage to reproduction in some species of birds and other animals is well known. Pollution fighters are carrying on the work of Rachel Carson, who first pointed out in her book, THE SILENT SPRING, the far-reaching effects of biocides. Some state governments and federal agencies, including the United States Department of the Interior, have stopped the use of DDT on their lands.

While there is evidence that DDT is a widespread harmful pollutant, not all pesticides can be banned in a crowded world that grows hungrier day by day. The green revolution cannot keep pace with the population explosion. Research in safer ways of controlling insects is being done in many areas, but long-term effects are difficult to determine. You can help by encouraging funds for research.

Know the difference between biocides that do not disappear quickly and those which are almost indestructible. DDT will stay on the earth far longer than the animals which it killed. To make matters worse, it concentrates inside certain organisms, and as it goes through the food chain it becomes much stronger. For example, DDT was put into a lake to destroy some insects that were destructive and annoying. It quickly disappeared from the water, for it was absorbed by the small plants and animals that lived there. Each one of these contained only a very minute part of DDT, about five parts per million, but it was several hundred times more concentrated than the DDT had been in the water.

Small fish in the lake ate millions of these tiny plants and animals in order to supply their bodies with food. Along with the plants and animals, they ate the DDT, which was part of them. Because of the large quantity that they consumed, the DDT became even more concentrated in their bodies.

As in a typical food chain, large fish ate the small fish, and DDT became even more concentrated. In some of the larger fish there were more than 2,000 parts per million. Birds which ate the fish absorbed the pesticide, and in a period of several years many of the birds died—some as long as five years after the last spraying of DDT, long after any trace of this pesticide could be found in the water.

Unfortunately, the earth is not vast enough to prevent worldwide contamination, for pesticides such as

DDT are released into the environment and travel from one end of the earth to the other. Because of unpredictable ways of concentration in animals, in air that moves over the world, and water that moves through the oceans of the world, the worldwide picture must be considered.

Action spurred by conservationists and organizations that are concerned with the environment has already had some effect. Three United States manufacturers announced that they would stop making DDT, and these companies in the past accounted for about 50 percent of this pesticide manufactured in the United States. Although the controversy still rages, people are looking harder than ever before at the long-term effect of pollutant and relating toxic substances in ecological cycles.

In developing countries, the attitude toward pesticides such as DDT is far different from that in the United States. Poor and hungry people have less sympathy for danger to wildlife and even possible long-term harm to human beings, when DDT costs about 15 cents a pound and other less harmful chemicals, more than a dollar a pound. Where DDT can prevent malaria, this chemical does not look like an enemy.

BEWARE OF POLLUTION FIGHTERS WHO CONSIDER ONLY ONE ASPECT OF THE PROBLEM.

As a pollution fighter, it is difficult to distinguish between the voices of reason and alarm. The situation is so serious that it is hard to distinguish between statements based on emotion and those based on fact. For example, you may read that the earth is running out of oxygen rapidly because of the polluted environment.

Careful measurements show that the amount of oxygen in the atmosphere near the earth has not changed during the past sixty years. This does not mean that pollution cannot affect the oxygen content of the atmosphere, but it does mean that sensible action must be based on reliable information.

BEWARE OF POLLUTION FIGHTERS WHO OVERREACT.

Too much emphasis on emotional appeal and too little work will not solve a pollution problem.

Some pollution fighters describe the crowding of Planet Earth in such terms that one pictures the Gobi Desert of a few years from now, with cars bumper to bumper. Their passengers are destroying the wilderness or looking for some open spaces that might be left on the earth. If you have visited a national park recently, you may be tempted to believe this humorous kind of exaggeration. It is worth bringing home the point that the wilderness is rapidly disappearing.

At the same time that the planners are encouraging people to spread out from the cities to avoid crowding, congestion, and other medical problems, other planners are fighting for wilderness areas to be saved for future generations. Although it may seem that these two groups are at odds with each other, such is not really the case.

BEWARE OF JOINING A BANDWAGON BEFORE EXPLORING THE FACTS.

INFORMATION IS ONE OF THE BEST WEAPONS OF A POLLUTION FIGHTER.

Some people say that the present ecology movement is a phase. "People in the United States have always been victors over their problems, and with technology and science that can send man to the moon, why worry?" But the environmental problem is a matter of survival and time is running out.

BEWARE OF COMPLACENT PEOPLE WHO ARE UNWILLING TO BELIEVE THE SERIOUSNESS OF THE PROBLEM OF POLLUTION AND WHO THINK THAT PROBLEMS WILL SOLVE THEMSELVES.

Since ecology studies and pollution fighting appeal to most intelligent people as a front for action, many politicians are promising real action in the area of environmental problems. Many are sincere; many are not.

BEWARE OF PROMISES THAT ARE UNREALISTIC. HOLD YOUR REPRESENTATIVES TO THEIR PROMISES. COMMUNITY ACTION SUGGESTIONS IN THE NEXT CHAPTER MAY HELP YOU TO DO THIS.

14 SOME POLLUTION FIGHTERS IN

The list of ways in which you can fight pollution is long and varied. Many little things, put together, help to make a big one; and the voices of many young people can form a powerful chorus. Here are just a few examples of things that small groups of young people have accomplished.

The Patriot Vagabond Club of Independence High School, Charlotte, North Carolina, was originally a camping group. Today they have joined the millions of people who are trying to find ways of preserving vanishing species and wilderness. They have conducted environmental conservation conferences, and are taking part in a fight to preserve Smith Island, a small uninhabited island off the coast of North Carolina. This effort has been successful for this island has turned back to its natural state.

ACTION

A group of 180 students at a girls' school in Connecticut played an important part in protecting some coastal wetlands there against dumping and filling. This tidal marsh had been used by The Thomas School in Rowayton, Connecticut, as an outdoor classroom for biology studies. In the spring of 1969 they discovered that it was being filled by contractors who were planning to build on it. Although no law had been violated, the girls were concerned because this bog provided a spawning area for fish and shellfish and a haven for migrating water birds. As a result, they formed an organization called Protect Your Environment (P.Y.E.).

P.Y.E. sold bumper stickers, gave out lapel buttons, sent members of their organization to speak in front of numerous community groups, and aroused public attention to the problem of the vanishing marshlands. The marsh which they were trying to protect was lost, but as a result of their efforts, a law was drawn up and passed by the state House of Representatives and Senate, to protect other wetlands.

Perhaps you have taken part in cleaning a river, a bay area, a lake, or a pond in your neighborhood. Girl Scout Troop 127 of Hackensack, New Jersey, enlisted support of Girl and Boy Scouts and other people, both young and old, in cleaning a section of the Hackensack River. The job was a big one. They used power winches on trucks to haul out heavy items such as washing machines, baby coaches, fuel oil tanks, and an unbelievable variety of materials. They cleaned about seven miles of the river over a period of time, and pulled out sixty truck loads of assorted debris.

Ecology action groups have brought together polluters from various industrial plants, colleges, and civic organizations to cope with specific problems of air and water pollution. Project Soar (Save Our American Resources) began in February 1971 as the Boy Scouts of America Conservation Good Turn. Numerous activities by thousands of boys play a part. If you want more information, contact your local scout unit.

Similar stories of education and determination are numerous and the results are beginning to show in many small ways. In many cases, young people are the spearheads.

Although millions of people have become pollution-minded, it is still important that each individual become an *active* pollution fighter. So often people say, "Why doesn't someone do something about it?" You are that someone who can do something about pollution. One of the most effective campaigns is an intelligent word-of-mouth education. The conservation war will not be

ACTION

A group of 180 students at a girls' school in Connecticut played an important part in protecting some coastal wetlands there against dumping and filling. This tidal marsh had been used by The Thomas School in Rowayton, Connecticut, as an outdoor classroom for biology studies. In the spring of 1969 they discovered that it was being filled by contractors who were planning to build on it. Although no law had been violated, the girls were concerned because this bog provided a spawning area for fish and shellfish and a haven for migrating water birds. As a result, they formed an organization called Protect Your Environment (P.Y.E.).

P.Y.E. sold bumper stickers, gave out lapel buttons, sent members of their organization to speak in front of numerous community groups, and aroused public attention to the problem of the vanishing marshlands. The marsh which they were trying to protect was lost, but as a result of their efforts, a law was drawn up and passed by the state House of Representatives and Senate, to protect other wetlands.

Perhaps you have taken part in cleaning a river, a bay area, a lake, or a pond in your neighborhood. Girl Scout Troop 127 of Hackensack, New Jersey, enlisted support of Girl and Boy Scouts and other people, both young and old, in cleaning a section of the Hackensack River. The job was a big one. They used power winches on trucks to haul out heavy items such as washing machines, baby coaches, fuel oil tanks, and an unbelievable variety of materials. They cleaned about seven miles of the river over a period of time, and pulled out sixty truck loads of assorted debris.

Ecology action groups have brought together polluters from various industrial plants, colleges, and civic organizations to cope with specific problems of air and water pollution. Project Soar (Save Our American Resources) began in February 1971 as the Boy Scouts of America Conservation Good Turn. Numerous activities by thousands of boys play a part. If you want more information, contact your local scout unit.

Similar stories of education and determination are numerous and the results are beginning to show in many small ways. In many cases, young people are the spearheads.

Although millions of people have become pollution-minded, it is still important that each individual become an *active* pollution fighter. So often people say, "Why doesn't someone do something about it?" You are that someone who can do something about pollution. One of the most effective campaigns is an intelligent word-of-mouth education. The conservation war will not be won in a short time. Many of the battles will be long and drawn out; but a growing army of amateur environmentalists with staying power is needed.

ECOS began with a small group of graduate students who were concerned with developing an awareness within human beings of the balance of nature and our dependency on this balance for survival. The group has grown rapidly and chapters are forming in different parts of the country. You may wish to contact ECOS, P.O. Box 1055, Chapel Hill, North Carolina 27514, for information about "ECO-KITS." This group is emphasizing the need for work over a long period of time. They quote an observer who said, "If the environmental crisis is a fad . . . it will probably be the last fad we ever have."

The Environment Action Bulletin is published weekly by Rodale Press, Inc., Emmaus, Pa. 18049. If you wish to keep informed on what small and large groups are doing currently, your librarian may subscribe to this for you. Many of the ideas may help with your own action programs.

The battle is just beginning, and time is short. In the following chapter you will find many ideas for further action.

15 SUGGESTIONS FOR FURTHER

The challenge of fighting pollution is inspiring, exciting, and vast. You can do much as an individual by following some of the suggestions in this book and in books listed on the following pages. You can work as one of a group with a local organization. You can get help from federal, state, and local agencies and from private organizations. Some of the most useful publications and organizations are listed below.

Many of the books have suggestions for organizing projects and tell of what others are doing in various areas. You may be able to use some of their experiences as models for your own fight against pollution.

Ask your librarian to obtain the following catalogues from conservation groups. These will lead you to many local action organizations, clearing houses, and service groups.

ACTION

The Conservation Directory. National Wildlife Federation, 1412 16th St. NW, Washington, D.C. 20036. Annually. Lists agencies, organizations, and officials concerned with natural resources, land use, and management.

State and Regional Conservation Councils. The Conservation Foundation, 1250 Connecticut Ave. NW, Washington, D.C. 20036. Lists of action organizations in various states and regions of the United States.

Community Action Agencies Atlas. Office of Economic Opportunity, Executive Office of the President, Washington, D.C. 20506. Lists community action agencies.

Catalogue of Federal Domestic Assistance. Information Center, Office of Economic Opportunity, Washington, D.C. 20506.

Community Action for Environmental Quality. Superintendent for Documents, U.S. Government Printing Office, Washington, D.C. 20402.

The following federal agencies publish many helpful guides and pamphlets which are either free or inexpensive. Be sure to specify the kind of information you want, or ask for a list of publications if you are not certain.

Bureau of Outdoor Recreation. Department of the Interior, Washington, D.C. 20240.

Federal Water Pollution Control Administration. Department of the Interior, 633 Indiana Ave. NW, Washington, D.C. 20240.

Department of Agriculture. Washington, D.C. 20250.

Look in your phone book under U.S. Government, Agriculture or County Extension Agents. Ask for information about various programs in your local area.

If you are working on a soil project, write to the Soil Conservation Service, Department of Agriculture. (Address above.)

If you are interested in recreational development of private woodlands, consult the Forest Service under the Department of Agriculture. (Address above.)

Department of Health, Education and Welfare, Washington, D.C. 20202.

Environmental Health Service, Park Lawn Bldg., 5600 Fishers Lane, Rockville, Md. 20852.

The Environmental Education Studies Staff may furnish a list of contacts which may assist you. Write 7th and D Streets SW, Washington, D.C. 20202.

Department of Housing and Urban Development, Washington, D.C. 20410.

Information about open space and outdoor recreation plans.

HOW TO ORGANIZE A COMMUNITY AWARENESS ACTION PROGRAM

First essential—You must be convinced that this is a real crisis, so that your sincerity will be contagious and will enable you to win the support of (1) school administrators, (2) faculty, and (3) student body.

Second essential—Students and teachers who are willing to use their imagination to come up with fresh ideas on how to educate the adult community, and willing to put their time and energies into achieving this objective.

Some Suggestions—The following are *some* of the things which you might initiate.

1 Identification of local, state, and national environmental problems.

2 A flow of letters to the editors of your local press offering solutions to environmental problems.

3 Articles in your school newspaper analyzing the problems and informing the people.

4 Distribution of fact sheets on the problems in your area. You may already know many of the specific problems, but you can get more information from your city, county, and state government. Government officials at all levels need interested and informed people to help them make decisions and take action in the overall public interest. *If they are uncooperative, then they are part of the problem*, and the public needs to know this as well.

5 Invite local radio and TV to do interviews on your projects.

6 Set up public meetings to which you invite local, state, and federal representatives to discuss the issues and to hear your and other citizens' concern. Get the facts before you speak. Informed and concerned citizens do indeed impress politicians and public officials.

7 Find out what ways you can show public support for pollution control, adequate zoning, better planned cities and transportation developments, and better management of natural resources. *Some of the ways*—a) Petitions; b) Letters; c) Phone Calls; d) Public Hearings; e) Marches.

8 Organize effective speaking teams and make them available to meetings and organizations in your community—business and civic groups, service clubs, women's clubs, youth groups, church groups, other schools. Invite experts in natural resource management, scientists, teachers, and other informed citizens to be an integral part of the effort.

9 Communicate with every elected and appointed official who represents or is employed by your community, city, county, state and keep them constantly aware of your concern. *They will react.*

10 Support and assist other groups that are working for resource conservation and a quality environment for man and other creatures.

11 Do everything possible to enlist as many people as possible into your efforts. Give everyone something to do —at least get them to voice their concern and support for effective environmental planning, programs, and legislation.

12 Form a publicity committee. Put your best students on it.

Everything you do, and every reason you have for doing it should be shared with your community. Stress the importance of *action now to insure a future.*

Local Agencies

Consult your phone book. Consult your local health department for the names of agencies responsible for environmental pollution control.

National Private Organizations

Many of these groups have local branches which will help you. You may find the local address in the *Conservation Directory*, or look in your local phone book.

State Agencies

Although titles vary in different states, you may find the correct one listed under State Agencies in your phone book. A letter addressed to Air Pollution Control at your state capitol, or to Department of Conservation or Department of Water Pollution Control, state capitol, will reach the proper agency even though the title is not precise. State Universities may be helpful in supplying information about pollution control.

IN EACH STATE: WHERE THE POLLUTION CONTROL IS CENTERED

STATE	AIR POLLUTION AGENCY	WATER POLLUTION AGENCY
Alabama	Department of Public Health State Office Building Montgomery, Ala. 36104 Director: Arthur Beck	Water Improvement Commission State Office Building Montgomery, Ala. 36104 Tech. Secy.: Arthur Beck

STATE	AIR POLLUTION AGENCY	WATER POLLUTION AGENCY
Alaska	Dept. of Health & Welfare Alaska Office Building Juneau, Alaska 99801 Chief of Environmental Health: James Anderegg	same
Arizona	Environmental Health Service Department of Health Hayden Plaza West 4019 N. 33rd Ave. Phoenix, Ariz. 85017 Health Commr: George Spendlove	same
Arkansas	Pollution Control Commission 1100 Harrington Ave. Little Rock, Ark. 72202 Director: S. Ladd Davies	same
California	Air Resources Board 1400 10th St. Sacramento, Calif. 95814 Chairman: A. J. Haagen-Smit	Water Resources Control Board 1416 9th St. Sacramento, Cal. 95814 Chairman: Kerry Mulligan
Colorado	Department of Public Health 4210 E. 11th Ave. Denver, Colo. 80220 Director: Roy Cleere	
Connecticut	Air Pollution Control Section Environmental Health Services Div. 79 Elm St. Hartford, Conn. 06115 Chief: Louis Prouix, Jr.	Water Resources Commission Room 223 State Office Building 650 Main St. Hartford, Conn. 06115 Director: John Curry

STATE	AIR POLLUTION AGENCY	WATER POLLUTION AGENCY
Delaware	Air & Water Resources Commission Lockerman St. & Legislative Ave. P.O. Box 916 Dover, Del. 19901 Director: Austin Heller	same
District of Columbia	Department of Public Health 300 Indiana Ave. N.W. Washington, D.C. 20001 Director: Murray Grant	same
Florida	Dept. of Air & Water Pollution Control Tallahassee, Fla. 32301 Director: Vincent Patton	same
Georgia	Department of Public Health 47 Trinity Ave. S.W. Atlanta, Ga. 30334 Air Quality Control Director: William Hansell	Water Quality Control Board 47 Trinity Ave. S.W. Atlanta, Ga. 30334 Exec. Secy.: R. S. Howard, Jr.
Hawaii	Environmental Health Division Dept. of Health P.O. Box 3378 Honolulu, Hawaii 96801 Director: Walter Quisenberry	same
Idaho	Air Pollution Control Section Dept. of Health 512 W. State St. Boise, Idaho 83707 Director: Vaughn Anderson	Engineering & Sanitation Div. Dept. of Health P.O. Box 640 Boise, Idaho 83701

STATE	AIR POLLUTION AGENCY	WATER POLLUTION AGENCY
Illinois	Air Pollution Control Board 616 State Office Building Springfield, Ill. 62706 Secretary: C. W. Klassen	Sanitary Water Board State Office Building 400 S. Spring Street Springfield, Ill. 62706
Indiana	Air Pollution Control Board 1330 W. Michigan St. Indianapolis, Ind. 46206 Secretary: Perry Miller	Stream Pollution Control Board 1330 W. Michigan St. Indianapolis, Ind. 46206 Tech. Secy.: Blucher Poole
Iowa	Environmental Engineering Service Dept. of Health Lucas State Office Building Des Moines, Iowa 50319 Chief: Paul Howser	same
Kansas	Air Quality Conservation Comm. State Office Building Topeka, Kansas 66612 Chairman: James Clark	Environmental Health Services Dept. of Health Topeka Ave. at Tenth St. Topeka, Kansas 66612 Health Office: Hugh Dierker
Kentucky	Air Pollution Control Comm. 275 E. Main St. Frankfort, Ky. 40601 Secretary: Ralph Pickard	Water Pollution Control Commission 275 E. Main St. Frankfort, Ky. 40601
Louisiana	Air Control Commission P.O. Box 60630 New Orleans, La. 70160 Chairman: Andrew Hedmeg	Stream Control Commission P.O. Drawer FC University Station Baton Rouge, La. 70803 Exec. Secy.: R. A. Lafleur

STATE	AIR POLLUTION AGENCY	WATER POLLUTION AGENCY
Maine	Water and Air Environmental Improvement Commission State House Augusta, Me. 04330 Chairman: Donaldson Koons	same
Maryland	Environmental Health Serv. Dept. of Health 2305 N. Charles St. Baltimore, Md. 21218 Air Quality Control Chief: Jean Schueneman	Dept. of Water Resources State Office Building Annapolis, Md. 21401 Director: Paul McKee
Massachusetts	Division of Air Pollution and Radiological Health Dept. of Public Health 600 Washington St. Boston, Mass. 02111 Director: James Dallas	Division of Water Pollution Control 100 Cambridge Street Boston, Mass. 02202 Director: Thomas McMahon
Michigan	Div. of Occupational Health Dept. of Public Health 3500 N. Logan Lansing, Mich. 48914 Chief, Air Pollution Control: B. D. Bloomfield	Water Resources Comm. Steven T. Mason Building Lansing, Mich. 48926 Exec. Secy.: Ralph Purdy
Minnesota	Pollution Control Agency 717 Delaware St. S.E. Minneapolis, Minn. 55440 Exec. Dir.: John Badalich	same

STATE	AIR POLLUTION AGENCY	WATER POLLUTION AGENCY
Mississippi	Air and Water Pollution Control Commission P.O. Box 827 Jackson, Miss. 39205 Exec. Secy.: Robert Wright	same
Missouri	Air Conservation Comm. 112 W. High St. Jefferson City, Mo. 65101 Exec. Secy., E. F. Porter, Jr.	Water Pollution Board P.O. Box 154 Jefferson City, Mo. 65101 Exec. Secy.: Jack Smith
Montana	Division of Air Pollution Dept. of Health Cogswell Building Helena, Mont. 59601 Director: Benjamin Wake	Water Pollution Council Dept. of Health Laboratory Building Helena, Mont. 59601 Secretary: Claiborne Brinck
Nebraska	no agency	Water Pollution Control Council Dept. of Health Box 94757 State House Station Lincoln, Neb. 68509 Dir. Health: Lynn Thompson
Nevada	Bureau of Environmental Health Dept. of Health, Welfare and Rehabilitation 201 S. Fall St. Carson City, Nev. 89701 Director: James Williams	same

STATE	AIR POLLUTION AGENCY	WATER POLLUTION AGENCY
New Hampshire	Air Pollution Control Agency 61 S. Spring St. Concord, N.H. 03301 Director: Forrest Bumford	Water Supply and Pollution Control Commission 61 S. Spring St. Concord, N.H. 03301 Exec. Dir.: William Healy
New Jersey	Div. of Clean Air & Water Dept. of Health P.O. Box 1540 Trenton, N.J. 08625 Director: Richard Sullivan	same
New Mexico	Department of Health and Social Services P.O. Box 2348 Santa Fe, New Mexico 87501 Director: Larry Gordon	same
New York	Division of Air Resources Dept. of Health 41 State St. Albany, N.Y. 12207 Asst. Commr.: Alexander Rihm, Jr.	Division of Pure Waters Dept. of Health 84 Holland Ave. Albany, N.Y. 12208 Dep. Commr.: Dwight Metzler
North Carolina	Dept. of Water and Air Resources P.O. Box 9392 Raleigh, N.C. 27603 Director: George Pickett	same
North Dakota	Environmental Health & Engineering Services Dept. of Health Bismarck, N.D. 58501 Chief: W. Van Heavelen	same

STATE	AIR POLLUTION AGENCY	WATER POLLUTION AGENCY
Ohio	Division of Engineering Dept. of Health 450 E. Town St. Columbus, O. 43215 Chief: George Eagle	Water Pollution Control Board Dept. of Health P.O. Box 118 Columbus, Ohio 43216 Chairman: E. W. Arnold
Oklahoma	Environmental Health Service Dept. of Health 3400 N. Eastern Oklahoma City, Okla. 73111 Commissioner: A. B. Colyar	same
Oregon	State Sanitary Authority P.O. Box 231 Portland, Ore. 97207 Secretary: Kenneth Spies	same
Pennsylvania	Bureau of Environmental Health Dept. of Health P.O. Box 90 Harrisburg, Pa. 17120 Director: Wesley Gilbertson	same
Rhode Island	Dept. of Health 335 State Office Building Providence, R.I. 02903 Director: Joseph Cannon	same
South Carolina	Pollution Control Authority J. Marion Sims Building Columbia, S.C. 29201 Exec. Director: W. T. Linton	same

STATE	AIR POLLUTION AGENCY	WATER POLLUTION AGENCY
South Dakota	Div. of Sanitary Engineering Dept. of Health Pierre, S.D. 57501 Health Off.: G. J. Van Heuvelen	same
Tennessee	Stream Pollution Control Board Cordell Hull Building Sixth Avenue, North Nashville, Tenn. 37219 Acting Dir.: D. P. Roberts	Stream Pollution Control Board Cordell Hull Building Sixth Avenue, North Nashville, Tenn. 37219 Director: S. Leary Jones
Texas	Div. of Occupational Health and Radiation Control 1100 W. 49th St. Austin, Tex. 78756 Director: Charles Barden	Water Quality Board 1108 Lavaca Street Austin, Tex. 78701 Director: Gordon Fulcher
Utah	Air Quality Section Division of Health 44 Medical Dr. Salt Lake City, Utah 84113 Chief: Grant Winn	Water Pollution Control Committee 44 Medical Ave. Salt Lake City, Utah 84113 Exec. Secy.: Lynn Thatcher
Vermont	Industrial Hygiene Division Dept. of Health P.O. Box 607 32 Spaulding St. Barre, Vt. 05641 Director: Harry Ashe	Dept. of Water Resources State Office Building Montpelier, Vt. 05602
Virginia	Air Pollution Control Board Room 902 Ninth St. Office Building Richmond, Va. 23219 Exec. Secy.: Richard Arey	Water Control Board P.O. Box 829 Richmond, Va. 23230 Exec. Secy.: A H. Paessier

STATE	AIR POLLUTION AGENCY	WATER POLLUTION AGENCY
Washington	Air Pollution Control Board 1510 Smith Tower Seattle, Wash. 98104 Exec. Dir.: Robert Stockman	Water Pollution Control Commission P.O. Box 829 Olympia, Wash. 98501 Director: James Behlke
West Virginia	Air Pollution Control Comm. 4108 MacCorkle Ave. S.E. Charleston, W. Va. 25304 Director: Carl Bear, II	Division of Water Resources Dept. of Natural Resources 1201 Greenbrier St. East Charleston, W. Va. 25311 Chief: Edgar Henry
Wisconsin	Div. .of Environmental Protection Dept. of Natural Resources P.O. Box 450 Madison, Wisconsin 53701 Acting Adminstrator: Thomas Frangos	same
Wyoming	Div. of Industrial Hygiene Dept. of Public Health State Office Building Cheyenne, Wyo. 82001 Director: Robert Sundin	same

16 Career Information And Pollution

Since public concern for environmental quality is exploding, many new professional workers will be needed to carry out the crusade against pollution. Will you make pollution fighting your career? Perhaps you would like to learn what courses to take, where help is needed, and other information. Since this changes frequently, the following organizations may be of help to you in supplying current information. Some of the organizations listed earlier in this book may be helpful, too.

AIR POLLUTION CONTROL ASSOCIATION
4400 Fifth Avenue, Pittsburgh, Pa. 15213

AMERICAN INSTITUTE OF BIOLOGICAL SCIENCES
3900 Wisconsin Avenue, Washington, D.C. 20016

Control Information

AMERICAN INSTITUTE OF CROP ECOLOGY
809 Dale Drive, Silver Spring, Md. 20910

AMERICAN SOCIETY FOR MICROBIOLOGY
115 Huron View Boulevard, Ann Arbor, Mich. 48103

AMERICAN SOCIETY OF SANITARY ENGINEERING
228 Standard Building, Cleveland, Ohio 44113

CITIZENS FOR CLEAN AIR
598 Madison Avenue, New York, N.Y. 10022

ECOLOGICAL SOCIETY OF AMERICA RADIATION ECOLOGY SECTION
Oak Ridge National Lab, Oak Ridge, Tenn. 37831

ENVIRONMENTAL ENGINEERING INTERSOCIETY BOARD
P.O. Box 9728, Washington, D.C. 20016

FRIENDS OF NATURE, INC.
Brooksville, Maine 04617

INTERNATIONAL UNION FOR CONSERVATION OF NATURE AND NATURAL RESOURCES
2000 P. Street N.W., Washington, D.C. 20036

NATIONAL ASSOCIATION OF WASTE MATERIAL PRODUCERS
20 North Wacker Drive, Chicago, Ill. 60606

NATIONAL AUDUBON SOCIETY
1130 Fifth Avenue, New York, N.Y. 10028

NATIONAL ASSOCIATION OF SANITARIANS
1550 Lincoln Street, Denver, Colo. 80203

NATIONAL ASSOCIATION OF SOIL AND WATER CONSERVATION DISTRICTS
1025 Vermont Avenue, Washington, D.C. 20005

NATIONAL COUNCIL FOR STREAM IMPROVEMENT
103 Park Avenue, New York, N.Y. 10017

NATIONAL WATER PURIFICATION FOUNDATION
209 South LaSalle Street, Chicago, Ill. 60603

POPULATION CRISIS COMMITTEE
1730 K Street N.W., Suite 314, Washington, D.C. 20006

SOIL CONSERVATION SOCIETY OF AMERICA
7515 Northeast Ankeny Road, Ankeny, Iowa 50021

STUDENT CONSERVATION ASSOCIATION
Sagamore Hill National Historic Site, Oyster Bay, N.Y. 11771

WATER POLLUTION CONTROL FEDERATION
3900 Wisconsin Avenue, Washington, D.C. 20016

WILDLIFE DISEASE ASSOCIATION
333 North Michigan Avenue, Chicago, Ill. 60601

WILDLIFE SOCIETY
3900 Wisconsin Avenue Suite S-176, Washington, D.C. 20016

The following private organizations may help you with a special project or interest.

CAMPAIGN TO CHECK THE POPULATION EXPLOSION
50 East 42nd Street, New York, N.Y. 10007

COMMITTEE FOR ENVIRONMENTAL INFORMATION
438 Skinker Blvd., St. Louis, Mo. 63130

CONSERVATION FOUNDATION
1250 Connecticut Avenue N.W., Washington, D.C. 20036

FRIENDS OF THE EARTH (F.O.E.)
30 E. 42nd Street, New York, N.Y. 10017

INTERNATIONAL UNION FOR CONSERVATION OF NATURE AND NATURAL RESOURCES
2000 P Street N.W., Washington, D.C. 20006

IZAAK WALTON LEAGUE OF AMERICA
1326 Waukegan Road, Glenview, Ill. 60025
Conservation Office: 719 13th Street N.W., Washington, D.C. 20005

KEEP AMERICA BEAUTIFUL, INC.
99 Park Avenue, New York, N.Y. 10016

LEAGUE OF WOMEN VOTERS
1730 M Street, N.W. Washington, D.C. 20036

NATIONAL ASSOCIATION OF SOIL AND WATER CONSERVATION DISTRICTS
1025 Vermont Avenue N.W., Washington, D.C. 20005

NATIONAL AUDUBON SOCIETY
1130 Fifth Avenue, New York, N.Y. 10028

NATIONAL COUNCIL ON NOISE POLLUTION
1625 K Street N.W., Washington, D.C. 20006

NATIONAL PARKS ASSOCIATION
1701 18th Street N.W., Washington, D.C. 20009

NATIONAL RECREATION AND PARK ASSOCIATION
1700 Pennsylvania Avenue N.W., Washington, D.C. 20006

NATIONAL WILDLIFE FEDERATION
1412 16th Street N.W., Washington, D.C. 20036

NATURAL SCIENCE FOR YOUTH FOUNDATION
763 Silvermine Rd., New Canaan, Conn. 06840

NATURE CONSERVANCY
1522 K Street N.W., Washington, D.C. 20005

OPEN SPACE INSTITUTE
145 E. 52nd Street, New York, N.Y. 10022

PLANNED PARENTHOOD—WORLD POPULATION
515 Madison Avenue, New York, N.Y. 10022

POLLUTION FIGHTERS / TODAY'S HEALTH
535 North Dearborn Street, Chicago, Ill. 60602
Invites ideas for pollution fighting and offers to pass suggestions along to appropriate committees or public health agencies.

POPULATION CRISIS COMMITTEE
1730 K Street N.W., Washington, D.C. 20006

POPULATION REFERENCE BUREAU, INC.
1755 Massachusetts Avenue N.W., Washington, D.C. 20036

SCIENTISTS' INSTITUTE FOR PUBLIC INFORMATION
30 E. 68th Street, New York, N.Y. 10021

SIERRA CLUB
1050 Mills Tower, San Francisco, Cal. 94104

WILDERNESS SOCIETY
729 15th Street N.W., Washington, D.C. 20005

YOUNG WOMEN'S CHRISTIAN ASSOCIATION
600 Lexington Avenue, New York, N.Y. 10022

ZERO POPULATION GROWTH
367 State Street, Los Altos, Cal. 94022

Bibliography

HARDBACK BOOKS

Allen, Durward L., *The Life of Prairies and Plains.* New York, McGraw-Hill, 1967.

Amos, William, *The Life of the Pond.* New York, McGraw-Hill, 1967.

———, *The Life of the Seashore.* New York, McGraw-Hill, 1966.

Berrill, N. J., *The Life of the Ocean.* New York, McGraw-Hill, 1966.

Boyle, Robert H., *The Hudson River, A Natural and Unnatural History.* New York, W. W. Norton, 1969.

Brooks, Maurice, *The Life of the Mountains.* New York, McGraw-Hill, 1967.

Dubos, René, *So Human an Animal.* New York, Charles Scribner's Sons, 1968. (Pulitzer Prize winner).

Ehrlich, Paul R. and Ann H. Ehrlich, *Population Resources Environment: Issues in Human Ecology.* San Francisco, W. H. Freeman, 1970.

Farb, Peter and the Editors of *Life. Ecology.* New York, Time, Life Nature Library, 1963.

Fisher, James, Noel Simon, Jack Vincent, and others. *Wildlife in Danger.* New York, The Viking Press.

Fuller, R. Buckminster, *Operating Manual for Spaceship Earth.* Carbondale, Southern Illinois University Press, 1969.

Graham, Frank, Jr., *Since Silent Spring.* Boston, Houghton Mifflin Co., 1970.

Herber, Lewis, *Crisis in Our Cities.* Englewood Cliffs, N.J., Prentice Hall, 1965.

Laycock, George, *The Diligent Destroyers.* Garden City, Doubleday, 1970.

McCormick, Jack, *The Life of the Forest.* New York, McGraw-Hill, 1966.

McHale, John, *The Future of the Future.* New York, George Braziller, 1969.

McHarg, Ian, *Designed with Nature.* New York, Natural History Press, 1970.

Milne, Lorus and Margery Milne, *Patterns of Survival.* Englewood Cliffs, N.J., Prentice Hall, 1967.

Mohr, Charles E. and Thomas L. Poulson, *The Life of the Cave.* New York, McGraw-Hill, 1967.

Niering, William A., *The Life of the Marsh.* New York, McGraw-Hill, 1966.

Pingle, Laurence, *The Only Earth We Have.* Macmillan (Collier-Macmillan Ltd., London), 1969.

Stephen, David, and James Lockie, *Nature's Way.* New York, McGraw-Hill, 1969.

Sutton, Ann and Myron Sutton, *The Life of the Desert.* New York, McGraw-Hill, 1966.

Teal, John and Mildred Teal, *Life & Death of the Salt Marsh.* Atlantic-Little Brown, 1969.

Usinger, Robert L., *The Life of Rivers and Streams.* New York, McGraw-Hill, 1967.

PAPERBACK BOOKS

Abrahamson, Dean E., *Environmental Cost of Electrical Power*. SIPI, New York, 1970.

Berg, George G., *Water Pollution*. New York, SIPI, 1970.

Bernarde, Melvin A., *Our Precarious Habitat*. New York, W. W. Norton, 1970.

Carson, Rachel, *Silent Spring*. New York, Crest, 1969.

Commoner, Barry, *Science and Survival*. New York, The Viking Press, 1966.

Cook, Robert C. and J. Lecht, *People: An Introduction to the Study of Population*. Washington, D.C., Columbia Books, 1968.

Dahlsten, Donald L. and Robert ven den Bosch, and Committee. *Pesticides*. New York, SIPI, 1970. *Nuclear Explosives in Peacetime*. New York, Ballantine Books, 1970.

Ehrlich, Paul R., *The Population Bomb*. New York, Ballantine, 1968.

Goran, Morris, *Experimental Earth Sciences*. Indianapolis, Bobbs-Merrill, 1967.

Hafner, Everett M. and Committee, *Environmental Education*. New York, SIPI, 1970.

Helfrich, Harold W., Jr., *Environmental Crises: Man's Struggle to Live with Himself*. New Haven, Yale University Press, 1970.

Leinwand, Gerald, General Editor, *Air and Water Pollution*. New York, Washington Square Press, 1969.

Marx, Wesley, *The Frail Ocean*. New York, Ballantine, 1969.

McClintock, Michael, and Committee, *Environmental Effects of Weapons Technology*. New York, SIPI, 1970.

Meade, Margaret, *Hunger*. New York, SIPI, 1970.

Mitchell, John G. with Constance L. Stallings, *Ecotactics: A Sierra Club Handbook for Environment Activists*. New York, Pocket Books, 1970.

Nadler, Allen, M.D., and others, *Air Pollution.* New York, Scientists' Institute for Public Information, (SIPI), 1970.

Odom, Eugene P., *Ecology.* New York, Holt, Rinehart and Winston, 1963.

Rienow, Robert R. and Leona L. Rienow, *Moment in the Sun.* New York, Ballantine, 1969.

Schwartz, William, Editor, *Voices for the Wilderness.* New York, Ballantine Books, 1969.

Selye, Hans, M.D., *The Stress of Life.* New York, McGraw-Hill, 1956.

Shurcliff, William A., *SST. Sonic Boom Handbook.* New York, Ballantine Books, 1970.

Storer, John H., *Man in the Web of Life.* New York, Signet, 1968.

Weaver, Elbert C., Editor, *Scientific Experiments in Environmental Pollution.* New York, Holt, Rinehart and Winston, 1968.

PAMPHLETS

Air Pollution Injury to Vegetation. USDHEW—1970. Supt. of Documents U.S. Government Printing Office, Washington, D.C. 20402.

Clean Water—It's Up to You. Izaak Walton League, 1326 Waukegan Rd., Glenview, Illinois 60025.

Community Action Program for Water Pollution Control and *Community Action Program for Air Pollution Control.* National Association of Counties, 1001 Connecticut Ave. N.W., Washington, D.C. 20036.

County Action for Outdoor Recreation. National Association of Counties, 1001 Connecticut Ave. N.W., Washington, D.C. 20036.

How to Preserve Your Area for Its Natural Value. The Nature Conservancy, 2039 K St., N.W., Washington, D.C. 20006. 1962.

A Little About Lots. (How to make Vestpocket Parks on

vacant lots.) Park Association of New York City, 15 Gramercy Park South, New York, N.Y. 10003.

A Nature Center for Your Community. The National Audubon Society, 1130 5th, Avenue. New York, N.Y. 10028, 1962.

Open Space for Urban America. Ann Louise Strong. Office of Metropolitan Development, Department of Housing and Urban Development, Washington, D.C. 20410.

Pollution by Pesticides. The Conservation Foundation, 1250 Connecticut Avenue, N.W., Washington, D.C. 20036.

Signs Out of Control. California Roadside Council, 2636 Ocean Avenue, San Francisco, California 94132.

Solid Waste Management. National Association of Counties, 1001 Connecticut Avenue, N.W., Washington, D.C. 20036.

So You'd Like to Do Something About Water Pollution. League of Women Voters, 1730 M Street, N.W., Washington, D.C. 20026.

Workbook for Clean Air. The Conservation Foundation, 1250 Connecticut Avenue, N.W., Washington, D.C. 20036.

Youth Takes the Lead. Urban Research Corporation, 5464 South Shore Drive, Chicago, Ill. 60615.

NEWSLETTERS

The Environmental Action Bulletin. Rodale Press, Inc., Emmaus, Penn.

National Parks, The Environmental Journal. National Parks Association, 1701 Eighteenth Street, N.W., Washington, D.C. 20009.

Index

About the Author and Artist

Margaret O. Hyde is the author of an outstanding list of over twenty books for young readers and the winner of the 1961 "Thomas Alva Edison Award" for the best children's science book. She has also written scripts of ANIMAL SECRETS for NBC-TV. They are: *How the Mind Begins* and *Can Human Nature Be Changed?* Mrs. Hyde graduated from Beaver College and received her M.A. degree from Columbia University. She was science consultant at the Lincoln School of Teacher's College, Columbia University. She now resides in Nyack, New York with her husband.

Don Lynch is a landscape painter who has had a lifelong interest in ecology. He lives in Nyack, New York, where he enjoys nature sketching while hiking in the nearby mountains and Hudson Valley. Mr. Lynch attended the Art Students League, the Society of Illustrators, and studied under Harvey Dunn. He has worked with leading advertising agencies and publishers and presently teaches painting and drawing at the High School of Art and Design in New York. FAVORITE TALES FROM SHAKESPEARE, BRONZE BILLY, MIRRORS, and EXPLORING NEW YORK STATE are among the many books he has illustrated.

5 Hyde, Margaret O.

or pollution
ighters only